Polymeric carbons –
carbon fibre, glass and char

G. M. JENKINS

University College of Swansea

K. KAWAMURA

Nihon University, Japan

Cambridge University Press

Cambridge

London · New York · Melbourne

Published by the Syndics of the Cambridge University Press
The Pitt Building, Trumpington Street, Cambridge CB2 1RP
Bentley House, 200 Euston Road, London NW1 2DB
32 East 57th Street, New York, NY 10022, USA
296 Beaconsfield Parade, Middle Park, Melbourne 3206, Australia

© Cambridge University Press 1976

First published 1976

Printed in Great Britain by William Clowes & Sons Limited
London, Colchester and Beccles

Library of Congress cataloguing in publication data
Jenkins, Gwyn Morgan.
    Polymeric carbons – carbon fibre, glass and char.
    Bibliography: p.
    Includes index.
    1. Carbon fibres. I. Kawamura, Kiyoshi, 1942–
II. Title.
TA418.9.F5J46   620.1'93   74-16995
ISBN 0 521 20693 6

Contents

Acknowledgements

We wish to acknowledge the encouragement and labour of past and present members of our laboratory at Swansea who have helped towards our understanding of this new range of materials. Especially we would mention Dr D. B. Fischbach of the University of Washington, Dr M. K. Halpin of the National Bureau of Standards in Dublin, Dr R. B. Matthews of the Whiteshell Laboratories of the AECL, Manitoba, Canada, and S. Bale, the present Morganite Carbon Research Fellow at Swansea, A. Easton of the Morganite Carbon Company, Dr T. Takezawa of Nihon University and Dr J. Weaver of Dunlop Ltd.

We are also indebted to W. Ruland, S. Yamada, W. Watt, A. Parmee, J. C. Lewis and many others for long conversations in connection with the matters reported herein.

We would also wish to thank the National Coal Board of the United Kingdom for sponsoring the main part of the research work around which we were able to construct this book.

Thanks are extended to other sponsors for their help in sponsoring essential research work of a subsidiary nature: General Refractories, British Steel Corporation, Science Research Council, and The Rocket Propulsion Laboratories.

We are grateful to Professor R. W. Cahn of Sussex for encouraging us in this endeavour and for reading the original manuscript.

Finally we wish to thank Professor T. Tsuzuku of Nihon University, Tokyo, for bringing us together in the first place, and our wives for their patience and understanding.

1 Introduction

1.1 Definitions and terminology

When many high polymers are heated in an inert atmosphere to temperatures above 300 °C they lose much of their non-carbon content as gases and change to forms of carbon. This process is termed *carbonization*. If they pass into a liquid or tarry state immediately prior to carbonization the resultant carbon is termed a *coke*. If they do not pass through such an intermediate phase, the resultant product is termed a *char*. The process by which the non-carbon content is thereafter eliminated is not complete until heat-treatment temperatures exceed 1000 °C. Below this temperature, the materials have intermediate properties between polymer and carbon and are termed *pyropolymers*. Above this temperature the material is termed a polymer or *polymeric carbon*. We prefer the latter terminology because it underlines the essentially polymeric nature of the carbon and differentiates it from the cokes produced from fused polymers. It is this type of material which is the main subject of this book. We must, of course, consider it in relation to the starting material, or *polymeric precursor*, and in relation to other carbons.

1.2 Range of subject

The most common and the earliest known char is *charcoal* which is the carbon produced by carbonizing wood, originally consisting mainly of polymeric (cellulose) fibres bonded together with lignin. In contrast to cokes, the final morphology is very close to the original wood. It is used as an absorbent in purifying gases and liquids, a quality common to all polymeric carbons with suitable activation.

Cellulose and cellulose rayon fibres do not fuse or degrade during carbonization, but retain something of their original shape. They were thus used by Edison as filaments in the first electric lamps. Latterly, at the Union Carbide Laboratories, the same fibres have been pulled at high temperatures to orient the polymeric carbon chains and so produce high strength, high modulus fibres.

Other polymeric carbons such as those derived from furfuryl alcohol and phenolic resins are used as binder material and impregnants in

1

standard carbon and graphite bricks and so are essential components in electrodes, moderator blocks and brushes. They are not used usually as fillers; indeed, most electrode graphite is made from so-called 'soft' filler and 'soft' binder, where 'soft' is the terminology for graphitizable cokes, that is, those which soften sufficiently for easy machineability on heating to 2700 °C. Electrodes made from 'hard' filler and 'hard' binders are relatively rare. '*Hard*' is the term used for non-graphitizable chars, that is, those which do not soften sufficiently for easy machineability on heating to 2700 °C. With few exceptions the 'hard' chars are derived from polymers. Because of the high carbon yield for these chars, the char-producing resins are used to impregnate electrodes to achieve a higher density and less permeability after their final firing in the Acheson process.

Davidson and Losty (1963) developed a process by which masses of cellulose fibres were agglomerated and carbonized to form electrically conducting glassy carbon artefacts with well-nigh zero permeability to all gases, and the ability to withstand savage chemical attack and very high temperature. Others (Carbone Lorraine, Vitreous Carbons and Tokai Electrodes) have produced glassy carbon artefacts using a variety of starting materials, including phenolic and polyfurfuryl alcohol resins.

Early in the 1960s it was found by Shindo in Japan that it was possible to produce carbon fibres from polyacrylonitrile (PAN) fibres with good strength and stiffness. Meanwhile, at Farnborough, Watt found that if these fibres were pulled in air during an early stage of the process before carbonization, the fibres became highly anisotropic and the resultant carbon fibres were strong and stiff enough to be used for reinforcing high polymers. This has prompted an enormous rise in research in polymeric carbons – mainly directed to improving the PAN process.

Polymeric carbons with basically similar molecular structure may act as powerful adsorbents with enormous available internal surfaces, in the form of activated charcoal, or as effective barriers, impermeable even to helium, in the form of glassy carbon. They can exist as weak, friable charcoals or as some of the strongest and stiffest fibres known to man. They form a distinctive group of materials which need to be differentiated from the soft graphitic carbons. The concept is introduced in this book of a graphitic ribbon network structure for all polymeric carbons as opposed to the extensive graphitic sheets which must exist in graphitic carbons. This will act as a constant theme throughout each chapter.

In Swansea, we have developed techniques for producing polymeric carbon artefacts from phenolic resins in the shape of fibres and discs.

Accordingly, we shall draw extensively on our results by way of illustration of the whole range of both isotropic and anisotropic polymeric carbons.

1.3   The carbon atom and the nature of the carbon–carbon bond

The first fundamental step in our quest into the nature of polymeric carbons must lie in the nature of the carbon atom, the carbon–carbon bond and the ways in which carbon atoms can be connected together to form a solid material.

There are three possible configurations of the outer electrons when the carbon atom is bonded to others:

Tetrahedral or sp_3 state. In this state all four electrons are absorbed into four evenly spaced hybrid orbitals. The most probable positions for the surrounding four atoms would then be at points which form a tetrahedron with the carbon nucleus at the centre. The σ bonds formed with 4 neighbours will then be at $109° 28'$ to each other. There are no electrons available to form subsidiary π bonds.

An example of tetrahedral carbon atoms bonded together would be in the gas ethane (H_3C—CH_3) which is at the start of a homologous series culminating in the long chain polymer polyethylene, consisting of a chain of methylene (—CH_2—) groups. Such carbon atoms can be bonded in three dimensions to form the cubic diamond lattice, thus producing a hard crystalline solid.

Trigonal or sp_2 state. In this state three electrons are absorbed in a symmetrical hybridized orbital system. The most probable positions for the orbital axes are coplanar and mutually at $120°$. The extra electron is in the free p state and is available for forming a subsidiary π bond.

An example of carbon atoms bonded together in the trigonal state would be the gas ethylene (CH_2=CH_2). Another would be polyvinylene (CH=CH)$_n$ which consists of strings of such carbon atoms and is a semiconducting polymer because the electrons in π bonds are delocalized. Benzene (C_6H_6) is an example of a ring of 6 equispaced trigonal carbon atoms arranged in a hexagonal pattern.

All graphitic materials consist of extensive parallel sheets of such carbon atoms arranged in such a pattern. We shall show that polymeric carbons are also made up of carbon in this state but arranged to form networks of long, narrow, entwined graphitic ribbons.

In order to explain the fact that hydrogen can be eliminated from carbons by heat-treatment, leaving no unpaired electrons, Mrozowski

(1952) proposes that carbon can exist in the s_2p_2 state and this is stable at the edges of graphite sheets. The bond angle is not known but it is not thought to be very different from the $120°$ existing for the normal sp_2 carbon atoms within the sheet.

The nature of the carbon–carbon bond between two atoms in the sp_2 state when the bond is bent or twisted is open to conjecture. This strained bond is important, however, in establishing a connection between edge-on sheets of graphite held at an angle to each other. It thus represents a strong boundary restraint.

Digonal or sp state. This is the state with two symmetrical hybridized orbitals whose electrons are capable of being absorbed in a molecular orbital bonding system. The only possible arrangement for neighbouring carbon atoms is on either side of the nucleus, the σ bonds being colinear and the coordination number being 2. The remaining two electrons are in the free p state and are available for forming subsidiary π bonds.

Examples of molecules containing carbon atoms in this state are the gas acetylene ($HC\equiv CH$) and 'carbyne' ($C\equiv C)_n$ which consists of chains of sp carbon atoms.

The bond energy between carbon atoms and their separation depend on the number of electrons contributing. The bond energy between two carbon atoms in the sp_3 state is 83 kcal mole^{-1}, and the distance between is 1.54 Å. This is so in such widely differing molecules as ethane, polyethylene and diamond.

With carbon atoms in the sp_2 state the extra p electrons increase the bond energy and decrease the atomic spacing. These changes are attributed to an increase in 'percentage double bond character'. Thus the C–C distance in ethylene is 1.353 Å. There exists a range of possible interatomic distances from 1.54 to 1.35 Å, depending on the degree of participation by the extra electrons. The bond energy is increased to 147 kcal mole^{-1} for the case of ethylene.

With atoms in the sp state, there is a further increase in bond energy and reduction in interatomic distance.

There exists a variability in bond distance within a molecule containing carbon atoms in the sp_2 and sp states. This is important, in that carbon–carbon bond distances are not necessarily characteristic of the presence of carbon in any one of the possible states.

An important factor is the ability of carbon atoms to rotate about a C–C bond. If the partners are in the sp_3 state and the bond is purely σ in character, rotation is easy and a high degree of flexibility in molecular orientation is possible. If the bond has any π component, however, rotation is severely hampered and a much more rigid inflexible structure is inevitable.

The bonds between the molecular units (the 'intermolecular cohesion') can only be weak van der Waals forces since no electrons are left over to form strong primary bonds. The small molecules or units will therefore be gases and liquids at room temperature. Polyethylene chains can be sheared past each other and oriented easily by cold work at room temperature. Perfect graphite sheets slide over each other easily at room temperature and will do so even down to 10 K. Diamond incorporates just one extended unit and so is extremely hard. Polycrystalline graphites and carbon must also have covalent cross-links between graphite sheets in neighbouring crystals in order to exist as coherent materials of significant strength.

1.4   Structure of polymers

Polymers consist of long chains held together by intermolecular forces of different amounts and intensities depending on the nature of the groupings attached to the chains and their ability to lie parallel to each other to form crystalline regions.

Some polymers, such as atactic polystyrene, lack symmetry so that it becomes impossible for the chains to pack closely; they can only exist as polymeric glasses with no crystallinity whatsoever. The chains are arranged completely randomly in space and can, in fact, be described quite adequately by three-dimensional random walk theory. Above the *glass point*, such materials are rubberlike – a condition in which chain segments are free to move by thermal activation to allow the free chains to assume a limitless number of configurations between fixed points, which are generally physical entanglements in thermoplastic materials and chemical cross-links in thermosetting materials.

Others, such as polythene, with a symmetrical arrangement of hydrogen atoms, crystallize easily. Crystalline regions consist of straight chain segments of polymer lying parallel to each other and separated by a constant distance. They are thus highly anisotropic; the strong and stiff covalent bonds only affecting the strain response in one direction (along the chain length). The strength and stiffness in all other directions are governed only by weak intermolecular cohesion. In high tensile polymer fibres all the crystalline regions lie approximately parallel to each other with the chains lying parallel to the fibre axis, and thus show a high degree of *preferred orientation*. In isotropic polymers the orientation of these essentially anisotropic crystallites is completely random and there is continuity of C–C bonding from one crystallite to another, usually via an 'amorphous' zone.

In recent years it has been demonstrated that in many crystalline polymers, the polymer chains cluster together to form 'fibrils' or 'microfibrils'. The crystalline regions detected by X-rays are merely parts where fibrils happen to be drawn taut. The full extent of a fibril is therefore not revealed by X-rays but only by the new technique of high resolution electron imageing.

A high degree of preferred orientation of fibrils results in high tensile modulus and strength. In natural forms of cellulose these can be very high. DuPont have recently produced an organic fibre (PRD-49) with the same strength as good quality steel ($2800 \, MN \, m^{-2}$), half its stiffness ($140 \, GN \, m^{-2}$) and a fifth of its specific gravity (1.47). The stiffness is still only a half of that theoretically possible.

When such polymers are quenched from the 'melt', they assume the amorphous arrangement existing at the higher temperature. By annealing at an intermediate temperature the crystals may grow and, if they are deformed plastically at room temperature, the intermolecular forces will be overcome and anisotropy can be induced.

Intermolecular cohesion depends on the nature of the radical appendages to the carbon chain. It is very strong in the case of polyvinyl chloride because of the electrical dipole formed on the carbon–chlorine axis. Other examples of such groupings are the hydroxyl (—OH) and the nitrile (—CN). It is commonly observed that polymers with high intermolecular cohesion are more likely to leave a carbon residue on pyrolysis.

If no thermal degradation or rearrangement of the chains takes place during carbonization, the final polymeric carbon must have a structure similar to that of the precursor and so these features we have discussed will have relevance to the structure of the carbonaceous product.

1.5 Structure of carbons

As indicated previously, there are only two crystalline forms of carbon – graphite and diamond.

Graphite consists of sheets of carbon atoms in the sp_2 state, each sheet being stacked in a hexagonal *ABA* sequence above each other as shown in fig. 1. The bonds in the basal plane are extremely stiff and strong and so the modulus in *a*-directions is very high; the material can withstand temperatures of 3300 °C before breaking up by thermal degradation alone. The bonds between the planes are of the weak van der Waals type and so the crystal can be sheared and cleaved easily in the plane perpendicular to the *c*-axis even at very low temperatures.

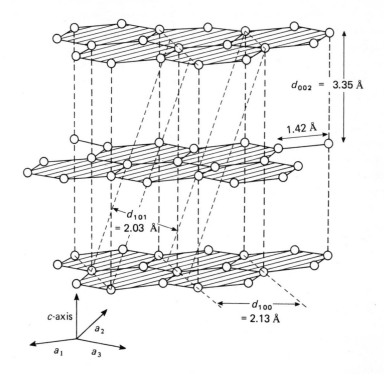

Fig. 1. Atomic structure of a perfect graphite crystal.

The distance between carbon atoms in the sheets is 1.42 Å while between perfect sheets the interlayer spacing is 3.354 Å. Graphite can be distorted permanently with ease by simply bending or shearing the sheets.

It is because the sheets can incorporate stable grown-in defects that carbons are stable up to very high temperatures before changing to graphite. In fact, '*graphitization*', defined as the establishment of a regular stacking of graphitic sheets, does not occur in '*graphitizable carbons*' until they are annealed above 2500 °C. The temperature range from 2500 to 3000 °C is called the '*graphitization temperature*' *range*. At this temperature it is presumed that vacancies in the graphite sheet become mobile enough to remove grown-in defects; graphitizable carbons become soft, as interplanar cohesion is reduced, and conduct both heat and electricity extremely well, as electron and phonon traps are removed. It is presumed that polymer carbons are a highly distorted form of graphite comprising ribbons rather than sheets, but this argument will be developed later.

When carbons are formed at low temperatures they contain many grown-in defects because thermal energy is not then sufficient to break carbon–carbon bonds, once formed. The presence of such defects increases interlamellar cohesion considerably and so such carbons are generally hard. The overall morphology of the graphite sheets is also laid down during carbonization. Simple annealing at high temperatures does not destroy this morphology.

Diamond consists of carbon atoms in the tetrahedral state bonded to accommodate all electrons without distortion, the carbon atoms fitting into the classical diamond type cubic lattice, consisting of two interpenetrating cubic (F) lattices based on 000 and $\frac{1}{4}\frac{1}{4}\frac{1}{4}$. Since all the bonds are equally strong and stiff, distortion is very difficult. It is possible to introduce dislocations and plastic flow at high temperatures but the structure will not tolerate the high degree of distortion possible in the graphite structure. Hence diamond is only found in the crystalline state and the lattice constants never vary. This contrasts with graphite in which the interlayer spacing can be varied easily and with the quasi-crystalline states of some carbons which are quite stable.

Because of the strong covalent bonding which prevents easy glide on all possible planes, diamond is hard and brittle. It reverts to graphite merely by heating above 1800 °C in an inert atmosphere. It is concluded that at such temperatures the sp_2 state is much more stable than the sp_3 state. Small distorted volumes of carbon in the sp_3 state will clearly revert at much lower temperatures than large perfect crystals.

Interrelationships between carbonaceous materials are summarized in fig. 2.

Hydrocarbon gases and liquids of low molecular weight can be pyrolysed in the gaseous phase to form tiny polyhedral carbon black particles, or pyrolysed on to a surface to produce slabs of pyrocarbon consisting of stacks of dimpled graphite sheets. When subjected to hot-working above 2500 °C, the dimples can be removed to form almost perfect slabs of graphite.

Graphite whiskers can be grown in a d.c. arc, typically with methane/argon mixtures under a pressure of 92 atmos at 2900 K. A solid boule of graphite is formed, comprising whiskers up to 3 cm in length and 10 μm in thickness, Bacon (1960). The whiskers are basal sheets rolled into scrolls containing a number of concentric elements. The whisker axis is parallel to the basal planes. Hence, whiskers are flexible, exhibit strengths greater than 20 $GN\,m^{-2}$, and Young's modulus in excess of 800 $GN\,m^{-2}$.

Hydrocarbons can be used to form high polymers and these are converted to chars or cokes on pyrolysis, unless chain scission takes

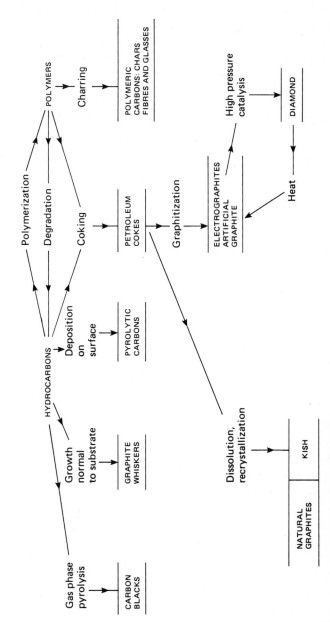

Fig. 2. The various forms of carbon.

place, whereupon only volatile products ensue. Generally, pure hydro-carbon polymers degrade into volatile products on pyrolysis. Polymeric systems containing oxygen or chlorine leave a carbon residue. Chars can be converted to soft graphite by catalytic graphitization. Soft cokes, of course, graphitize to form 'artificial' graphites or 'electro-graphites'. If electrographite is subjected to high pressures in the presence of a catalyst it changes to diamond. Diamond changes spontaneously at ordinary pressure to graphite above 1800 °C.

2 Pyrolysis of polymers

2.0 Generalities

When polymers are heated in an inert atmosphere reactions take place initially within the C–C chain. These intramolecular reactions result in one of three possibilities:

(i) The chains degrade into small molecules and the products are evolved as gases leaving little or no carbon behind. An example of this behaviour is shown by polyethylene.

(ii) The chains collapse to form aromatic lamellae and so move into a plastic phase from which the lamellae stack above each other to form spherulitic liquid crystals – the so-called 'mesophase'. The result is a crystalline anisotropic coke which on heating to 2700 °C will soften to a graphitic material. An example of this behaviour would be polyvinyl chloride.

(iii) The chains remain intact and merely coalesce with neighbours; the material does not pass through a plastic state. The final char is isotropic if the precursor material is isotropic and can be anisotropic if the precursor material is anisotropic. In both cases the material is not graphitizable when heated to 2700 °C and remains relatively hard. An example of a material which behaves like this is polyvinylidene chloride, which forms a 'hard' char.

If we use only pure polymers the carbons produced from cases (ii) and (iii) form distinct groups with very different properties. However, if we try to incorporate such carbons as carbon black (produced directly from the gas phase), composite materials with intermediate properties are apparent which makes the grouping difficult. In this present publication we shall ignore them.

The mechanisms of pyrolysis and carbonization are not understood because it is very difficult to determine the crystallographic and chemical structure of intermediates existing during carbonization. The explanations given to date for the mechanism of carbonization are far from giving us a scientific understanding of this complicated process.

In general, when polymers of type (iii) are pyrolysed the following stages of structural change are noted:

I. *The precarbonization stage.* The materials turn black at an early stage, usually at the very beginning of a regime of rapid weight loss, or

11

during pre-oxidation, when linear conjugated carbon–carbon systems start to form. All 'loose' molecules such as excess monomer or solvent are removed.

II. *The carbonization regime.* This is a regime of rapid weight loss, typically between 300 and 500 °C, in which oxygen, nitrogen, chlorine, etc. are removed. The background absorption in the infra-red gradually increases as the conjugated systems get longer. At any given temperature the process may be stopped, the time taken to achieve a weight loss characteristic of the temperature being of the order of minutes. This is ascribed to the increasing difficulty in eliminating molecules as strain energy is built up in a polymeric network with the increase in cross-linking between chains. The result is a loose network of linear conjugated systems which are still isolated electronically from each other.

III. *Dehydrogenation.* At the end of stage II, the material still contains, typically, one hydrogen for every two carbon atoms, which corresponds to the formula of an aromatic ladder polymer. Thus there exists a third regime between 500 and 1200 °C in which hydrogen is gradually eliminated, the hydrogen:carbon ratio being a characteristic of a given heat-treatment temperature.

The electrical conductivity increases rapidly as separate conjugated systems become interconnected to form a conducting network. The density, hardness and stiffness increase, the damping and permeability decrease.

IV. *Annealing.* Heat-treatment above 1200 °C is characterized by a sharpening of the X-ray lines as the component crystallites achieve greater perfection and defects are progressively removed. The material becomes slightly softer but still remains relatively hard and amorphous with respect to graphite materials even after heat treatment at 3000 °C.

In this chapter we shall concentrate on the chemistry of the first three stages. The final process will be more appropriately discussed in the chapter on structure.

The chemical structure of the starting material, or precursor, is the most important factor in determining the nature of carbonization. The thermal reactivity of an organic molecule depends on factors such as its size, the ease of free radical formation and the presence of substituents in aromatic rings. If one considers various types of aromatic compounds, fundamental molecular parameters such as the ionization potential correlate with thermal reactivity. On the other hand, there seems to be little obvious correlation between the structure of the starting organic molecule and the degree of crystallinity of the resultant carbon.

The initial chemical reactions are more important, and these are:

Bond cleavage at the most reactive molecular site to produce a free radical intermediate;

Rearrangement of radicals to more stable intermediates;

Polymerization of radical units;

Elimination of hydrogen from polymerized structures.

The situation is complicated by the fact that these reactions do not proceed in distinct steps but often occur simultaneously.

After studying the electron spin resonance of carbonaceous materials during carbonization, Singer (1968) pointed out that hydrogen transfer reactions are particularly important in the initial formation of radicals, which can then rearrange to form more stable intermediates. Properly controlled aggregation of the stable radical units can lead to quite perfect large aromatic systems but premature aggregation can lead to highly disordered structures such as those which occur in the formation of glassy carbon.

Initial chemical reactions in the starting materials can be controlled chemically and mechanically, and the structure and properties of ensuing carbons are influenced by the treatment imposed during carbonization.

For instance, Kipling *et al.* (1964) showed that polyvinyl chloride (PVC) oxidized below 200 °C in air gives rise to a non-graphitic carbon, whereas pyrolysis without pre-oxidation yields a 'soft' coke. Otani (1965) showed that carbon fibres made from PVC pitch are converted into an extremely non-graphitizing carbon after heavy oxidation of the pitch with ozone before carbonization. Otani oxidized extruded PVC pitch fibres in order to maintain the fibrous shape during carbonization because a graphitizable PVC pitch melts. This suggests that, on oxidation, oxygen bridges are formed between aromatic molecules which inhibit fusion during the later stages of carbonization. Together with the fact that carbonaceous materials which contain much oxygen are non-graphitizing, these results indicate that ether linkages (—C—O—C—) inhibit rearrangements to allow the growth of extensive aromatic crystallites. It is possible to convert all graphitizable polymeric materials into non-graphitizing carbons by oxidizing the starting materials in the early stages of carbonization.

Furthermore, it is found that the application of stresses during carbonization can influence the graphitizability of the resultant carbons. Noda (private communication) showed that polyvinyl chloride carbonized in an autoclave under high pressure produced a non-graphitizing carbon. However, interpretation of the results of carbonization in an autoclave is difficult because the presence of the escaping gases from organic substances can also influence the degree of crystallinity of the resultant carbons.

2.1   Carbon yield of polymers

A simple test is to heat a known weight of the polymer in an inert atmosphere to, say, 1000 °C and weigh the carbon residue. We can then determine the carbon yield which is the ratio of the weight of the carbon residue to the initial weight and the conversion efficiency which is the ratio of the weight of carbon in the residue to the weight of carbon in the original resin.

Typical examples are shown in table 1.

TABLE 1   *Carbon yields of various polymeric precursors*

Polymeric precursors	Carbon yield (%)	Comment
Coal tar pitches	50	High yield of coke
Petroleum fractions	Variable	
Polyvinyl chloride	42	
Phenol–formaldehyde	52	High yield of char
Epoxidized phenol–formaldehyde	50	
Phenol–benzaldehyde	37	
Oxidized polystyrene	55	
Polyfurfuryl alcohol	50	
Polyvinyl alcohol	50	
Polyacrylonitrile	44	
Polyvinylidene chloride	25	
Cellulose	20	
Polybutylene rubber	10	Moderate yield of char
Cellulose acetate		
Melamine formaldehyde		
Polyvinyl acetate		Negligible yields
Ethyl cellulose		
A typical epoxy resin		
Acrylonitrile–styrene copolymer		
Polystyrene	5	
Polyamides		
Polyisobutylene		
Polyethylene		
Polymethylmethacrylate		

Many long-chain polymers break down completely into gaseous products. Madorsky (1953) has shown that polyethylene at one extreme and polytetrafluoroethylene ($-CF_2 \cdot CF_2-$)$_n$ at the other leave no carbon residue. Polyvinyl fluoride ($-CH_2 \cdot CHF-$)$_n$ and polytrifluoroethylene ($-CHF \cdot CF_2-$)$_n$ also leave very little residue, but (CF_2-CH_2)$_n$ leaves nearly 30 % carbon residue by weight which accounts for all the carbon originally present in the polymer.

Winslow (1958) shows that polyethylene breaks down into gaseous products at 450–480 °C, but is quite stable up to that point. Polystyrene also breaks down completely, but at a much lower temperature (370–420 °C).

When polyvinyl chloride ($CH_2 \cdot CHCl$)$_n$ is heated it starts decomposing at 240 °C with evolution of HCl, leaving behind nearly all its original carbon (96.2 %) in the form of coke. Polyvinylidene chloride ($CH_2 \cdot CCl_2$)$_n$ similarly gives off HCl, leaving behind almost all of its carbon. Poly-acrylonitrile ($CH_2 \cdot CH(CN)$)$_n$ leaves a char consisting of most of the original carbon with elimination of only nitrogen and hydrogen. Poly-vinyl alcohol ($CH_2-CH(OH)$)$_n$ leaves most of its carbon behind on pyrolysis.

Winslow (1958) made a polystyrene-type polymer from a monomer consisting of 56 % *m*-divinyl benzene, 40 % *m*-ethylvinyl benzene 40 % and 4 % *m*-diethyl benzene. In order to stabilize it, the copolymer was heated in air at 250 °C. This preoxidized copolymer passed through a viscous phase and turned completely black only on pyrolyzing to 450 °C. The weight loss was only 45 %. Thus oxidation of polymers can prevent chain breakdown into gaseous products. Other polymers (Kipling *et al.*, 1964) also change their carbonization characteristics after pre-oxidation. Polyvinyl chloride, for instance, leaves a char instead of a coke in the case of untreated polymer.

Mackay (1969) has recently determined the carbon-coke yield for a whole range of polymers. He comes to the conclusion that carbon yield cannot be predicted in many cases. Linear polymers such as polystyrene or poly-*p*-xylene, even though having a high degree of aromaticity, yield low carbon char percentages. Linear polymers of furfuryl alcohol and polyphenylene oxide with aromatic groups interspersed with ether linkages provide good carbon yields. Conversion efficiency for polyfurfuryl alcohol, for example, was a high 90.5 %. Polyphenylene oxide has a carbon yield of 56 %.

Carbon yield does not seem to be influenced by whether the polymer is thermoplastic or thermosetting, linear or cross-linked, but whether it is capable of cyclization, ring fusion or chain coalescence at the onset of carbonization. In general, the resins should have a high degree of aromaticity and high molecular weight. There should be no more than one carbon atom between aromatic rings, because chain scissions will take place leading to volatilization of the fragmented parts – witnessed in a low yield from poly-*p*-xylene.

Nitrogen should be in the ring structure and not in the chain. For example, polybenzimidazole is more stable and provides higher carbon yields than amine cured epoxy resins. Other elements such as sulphur do not affect the stability of the polymer, but do result in lower char yields and low conversion efficiency.

All phenolic polymers except those in which the *para* position is blocked by either a phenyl or methyl group are good char-forming materials. Phenol–furfuraldehyde has a high carbon yield (62 %) and conversion efficiency (80 %), higher than phenol–formaldehyde (76 %).

Resorcinol–, *p-p*-dihydroxybiphenyl–, and 1,5-naphthalenediol– formaldehyde all have conversion efficiencies of 80 %. The last-mentioned does not show any tendency to melt or flow during the carbonization process. Only *p*-phenyl phenol–formaldehyde has a low conversion efficiency of 10 %.

Epoxy resins, in which R—CH$_2$—CH—CH$_2$ is cured with a suitable agent, do not in general have a high carbon yield. Exceptions are epoxidized phenol–formaldehyde cured with boron trifluoride amine which provides a 50 % carbon yield and phenol–glyoxaldehyde resin with 54 %. These are excellent materials for preparing carbon–carbon composite structures in that the good adhesion before carbonization is partly retained.

2.2 Detailed chemistry of pyrolysis: specific pyrolyses

Data for the determination of the chemistry of pyrolysis are provided by thermo-gravimetric analysis (TGA), thermo-volumetric analysis (TVA), differential thermal analysis (DTA), analysis of gaseous products, and by arresting the pyrolysis at various stages to analyse the residue. The techniques and the results on the thermal degradation of organic polymers are described by Madorsky (1964).

In DTA, thermograms of the pyrolysing material are drawn which represent small differences in temperature between the sample and an inert reference. The temperature at which heat is absorbed or given out is identified and, of course, the processes identified as endothermic or exothermic. However, the interpretation of a thermogram can only be qualitative and does not enable us to resolve the many simultaneous processes.

Gas phase chromatography using solid absorbents in packed columns is important in the analysis of gaseous products of the pyrolysis of organic polymers. Mass spectrometry is the fastest and most accurate method for analysing light hydrocarbons, for instance. Only a few milligrams of material is required, enabling the very small fractions from gas chromatography to be identified.

Analysis of the complicated residue is normally determined with reference to its infra-red absorption spectrum. The residue is crushed and mixed with KBr, which does not have an infra-red spectrum clashing

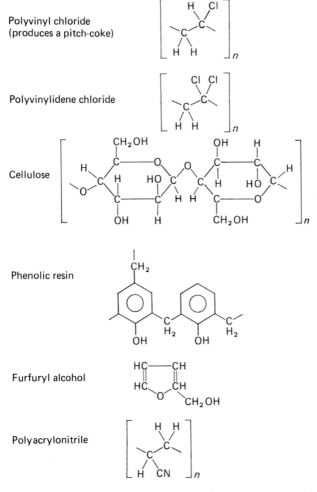

Fig. 3. Structures of polymeric systems which leave chars on carbonization.

with organic materials, and compacted to form thin pellets. For identification work quite an extensive library needs to be built up of infra-red absorption spectra of related pure compounds.

Magnetic and electrical measurements on the carbonaceous residue are useful. Unpaired electrons are attributable to broken bonds and these can be detected in air by means of a modulation spectrometer for measuring the paramagnetic resonance absorption (Winslow, 1956). Measurements of d.c. resistivity can be obtained by finely grinding the material and packing it firmly between brass electrodes in a non-conducting die.

X-ray measurements are used to try to characterize the carbonization process, but the patterns are so complex and non-crystalline that it is difficult to work out exact structure changes.

The number of polymers it is possible to carbonize is clearly enormous. Work on the detailed chemistry of carbonization has concentrated on a few which have been chosen for availability, high yield and simplicity of structure; we shall consider these in detail in this section. The structures are illustrated in fig. 3 in a simplified way.

2.2.1   Pyrolysis of polyvinyl and polyvinylidene chloride

Polyvinyl chloride (PVC) consists of a chain of sp_3 carbon atoms with a chlorine atom attached to alternate carbon atoms, $(CH_2—CH \cdot Cl)_n$. The remaining carbon atoms are clothed with hydrogen. It is formed by the addition polymerization of the monomer, vinyl chloride. The chlorine atom attracts electrons from the carbon chain and so produces a polar group. The intermolecular cohesion is greater than the simple van der Waals forces existing in polyethylene. It thus has a higher softening point (120 °C) and a greater stiffness. In order to reduce the intermolecular cohesion so as to increase the strain to fracture and improve mouldability, plasticizers are added. These are essentially polar organic solvents with a high boiling point which wet the long chains to separate them.

Polyvinylidene chloride (PVDC) consists of a sp_3 carbon chain with two chlorine atoms attached to every other atom, $(CH_2—CCl_2)_n$. Because of the greater symmetry, this material is highly crystalline, in that the long chains tend to be arranged in a linear fashion and parallel to each other. Because of this crystallinity the average intermolecular force increases and the material has a higher softening point than PVC (130 °C). It is also even more brittle and much more difficult to work. It is therefore usually produced as a *copolymer* with PVC, manufactured under the name 'Saran', produced by polymerizing comonomer mixtures of vinyl chloride and vinylidene chloride. It consists of a chain of mixed *mer* units with a certain proportion of VC mer units in a chain consisting mainly of VDC mer units. Its softening point (105 °C) is less than that of either PVC or PVDC.

Particular interest has been taken in the comparison of the pyrolyses of PVC and PVDC which differ by only one chlorine atom per unit yet pyrolyse in quite different fashions, one leaving a graphitizable, the other a non-graphitizable carbon residue. The detailed studies of their pyrolysis has been carried out especially by loss of weight (Kipling and McEnaney, 1966), by the analysis of the gases evolved (Gilbert and Kipling, 1962), by the examination of the residue with X-rays (Badami and Hussey, 1963), by the examination of the residue by infra-red spectroscopy, by the measurement of the paramagnetic resonance in the residue and by the density (Dacey and Cadenhead, 1960).

When PVC is heated in the absence of oxygen above 200 °C HCl is evolved and the material darkens to form a polymer presumed to have the structure $(CH{=}CH)_n$. On heating to 380 °C the polymer becomes very fluid due to thermal degradation. The material solidifies as it is carbonized to form a coke. It is inferred that the polymer chain breaks down and the component parts cyclize to form small discrete aromatic products which coalesce to form larger aromatic sheets which stack one above the other to form a liquid crystal. The liquid crystals combine to form large grains which become solid and establish the grain boundaries of the final graphitized coke. Polished sections of this coke examined optically show it to be highly crystalline, containing a mass of anisotropic crystallites even when formed at 500 °C. Graphitization yields a soft coke but does not change the crystal morphology.

PVDC decomposes with the evolution of HCl at temperatures as low as 100 °C but this is most marked between 140 and 200 °C. The decomposition tends towards a limit corresponding to a loss of one molecule of HCl per (CH_2CCl_2)mer unit. Further decomposition, however, is detectable if the temperature is maintained for some time. If the temperature is raised 50 °C, rapid further decomposition occurs to a new plateau. Thus successive temperature increments bring about a stepwise decomposition curve.

The mechanism for this decomposition is, according to Everett and Redman (1963), the loss of 1 HCl molecule per pair of carbon atoms, leading to the formation of a polyene chain:

$$(CH_2{-}CCl_2)_n \rightarrow (CH{=}CCl)_n + nHCl$$

This is a first-order reaction with a relatively high activation energy of 50 kcal mole^{-1} and involves a change in the configuration of the polymeric chains because of the changed valency angles, so that the reaction could not occur without diffusion of the chains in the solid state.

Further loss of HCl cannot occur by an intramolecular mechanism because loss of HCl would give rise to linear polyacetylene chains and the internal reorganization into straight chains is highly improbable. Coalescence of neighbouring chains, probably by a Diels–Alder type of reaction, must occur before further HCl can be eliminated. This is a second-order reaction with a small frequency factor (10^4 s^{-1} mole^{-1}) and a small activation energy.

Further coalescence can occur with other parts of the same chain, or, competitively, with another adjacent chain. Space models show that this cross-linking reaction has a complicated stereochemistry and so the kinetics must be complicated.

To account for the stepwise decomposition, it is necessary to assume that the cross-linking reaction rate constant falls off rapidly as the degree of cross-linking increases. This is because competition between chains in coalescing with a common neighbouring chain, for instance, must increase the strain energy of the system.

2.2.2   Cellulose polymers

These are all based on natural cellulose, which is the most common natural polymer, occurring as the stiffening component in all plants – it is the main constituent of wood and occurs in a pure form in the cotton plant. It is composed of anhydroglucose groups linked together by oxygen atoms to form a long stiff chain as shown in fig. 3.

It is estimated that natural cellulose contains chains with between 3000 and 4000 glucose units. The presence of numerous polar hydroxyl groups and a high molecular weight result in very strong intermolecular forces between the chains. These give high rigidity and crystallinity, resulting in a high softening point and insolubility; cellulose decomposes before it begins to soften. To make cellulose more amenable to manufacturing operations, it is necessary to decrease these intermolecular forces by reducing the molecular weight and/or by neutralizing the polar nature of the hydroxyl groups, for example, in regenerated cellulose (viscose) and cellulose acetate.

Regenerated cellulose has a much smaller molecular weight (300–500 glucose units). This lower weight greatly reduces the intermolecular forces, thereby making it possible to obtain cellulose in solution. The reduction of molecular weight is accomplished by treating cellulose pulp with sodium hydroxide solution to make soda cellulose. This is followed by ageing the soda cellulose by oxidation which reduces the degree of polymerization. After ageing, the cellulose is churned with carbon disulphide to form a gelatinous mass of sodium cellulose xanthate. This is stirred with dilute caustic soda solution to form a liquid known as 'viscose', which is ripened and spun into an acid coagulating bath. Normal additions are glucose, to increase pliability and softness, and zinc sulphate, for added strength.

The pyrolysis of cellulose has been studied by Losty and Blakelock (1965). On heating cellulose to 250 °C there is a 40% weight loss with the removal of —OH and —CH$_2$OH side groups. The free valencies thus created are probably satisfied by the chemisorption of hydrogen since electron spin resonance measurements fail to detect a significant proportion of free radicals. The next stage in the pyrolysis involves the removal of the oxygen from the pyranose ring and the glucosidic link of the cellulose molecule with the liberation of CO$_2$, CO and H$_2$O and the formation of some tar. Infra-red spectroscopy suggests that the pyranose ring is broken across the oxygen linkage to form a —COOH group. This process is gradual over the heat-treatment temperature range 240–370 °C, corresponding to 40–60% weight loss, and is really the homogeneous nucleation of small aromatic structures by the partial removal of O and H within the polymer chain causing it to straighten somewhat. Between 400 and 550 °C, the number of free spins increases, but the line width decreases, implying an increasing number of free radicals which are capable of interaction to sharpen the resonance line. For every 50 carbon atoms, there are 32 hydrogen atoms at 370 °C and 20 at 550 °C. It is assumed, therefore, that in this temperature range the modified cellulose chains progressively coalesce, the degree of coalescence depending, because of competition between neighbouring chains, on the heat-treatment temperature. Between 550 and 750 °C the hydrogen content falls to 5 hydrogens per 50 carbons, with a decrease in the number of free spins, due probably to a further coalescence. However, the maximum number of free spins observed represents only one spin for every 500 carbon atoms.

Tang and Bacon (1964) studied the low temperature pyrolysis of rayon fibre, measuring weight loss, dimensional shrinkage, chemical analysis, X-ray diffraction and infra-red spectroscopy. They proposed the following four stages of carbonization: I, physical desorption of water (25–150 °C); II, dehydration from the cellulose unit (150–240 °C); III, thermal cleavage of the glucosidic linkage, and scission of other C–O bonds and some C–C bonds via free radical reaction (240–400 °C); IV, aromatization (400 °C and above). After studying the Young's modulus and stress–strain curves of cellulose and partially dehydrated cellulose fibres, they concluded that the dehydration process in stage II is essentially intramolecular rather than intermolecular. They suggested that the final breakdown of each cellulose ring unit in stage III would result in a residue containing four carbon atoms which then represent the 'building blocks' for repolymerization into a 'carbon polymer' and, ultimately, a graphite structure. They considered that the degradation and repolymerization process in stages III and IV was a chain reaction via a free radical mechanism.

Pure cellulose in the form of rayon fibre has been used successfully as a precursor for high modulus carbon fibre, as will be described in section 3.2.2. In the form of cellulose pulp it is a suitable precursor for glassy carbon ware, as will be described in section 3.1.1.

When wood is pyrolysed, many changes take place consecutively. Optical and scanning electron microscope observation shows that cellular integrity is maintained even in large pieces of wood provided the temperature rise is controlled to a low rate. In charred American cherry (Blankenhorn, private communication), both upright and procumbent ray cells can be seen, and careful inspection reveals films, longitudinal parenchyma, helical thickening, simple perforation plates and vessel pits, all intact. There is an initial large drop in stiffness and strength with large weight loss and shrinkage between 200 and 500 °C. Heat-treatment above 500 °C produces further weight loss as hydrogen is evolved, with marked densification. The strength and stiffness increases remarkably to approach the values of the original wood.

2.2.3   Phenolic resins

These constitute an important group of polymers in which the chain consists of a phenolic group interspersed with, for instance, a methylene group. They are formed by a condensation reaction between, for instance, phenol and formaldehyde, with elimination of water. If excess formaldehyde is present, methylene cross-links between chains are possible, also with elimination of water.

Phenolic resins can be produced by one-stage and two-stage processes. In the one-stage process, phenol is reacted with excess formaldehyde, so that the phenol-to-formaldehyde (P:F) ratio is less than one. The mixture is heated in the presence of alkaline catalysts such as sodium hydroxide or ammonia. The reaction is interrupted at an early stage, so that the actual degree of polycondensation is carried out to either the A- or B-stage resins. The A-stage resin, called 'Resol', is a relatively short, low molecular weight, linear polymer; it is completely soluble in the alkaline solution. The B-stage resin, called 'Resitol, is a rather long linear polymer with a slight amount of cross-linking between chains. It is insoluble in alkaline solution, but readily soluble in organic solvents. When cooled, the resin becomes hard and brittle, but on heating, becomes soft again. A- and B-resins are used for adhesives, castings, plastics and laminates. Heating to a higher temperature causes extensive cross-linking and the formation of a hard and rigid solid, infusible and insoluble – the so-called C-stage resin.

In a two-stage process, less formaldehyde is initially introduced so that the P:F ratio is larger than one. The material is heated in the presence of an acid catalyst, and the reaction is allowed to proceed until no further chemical changes occur. The resulting resin is called 'Novolac' and is fusible and soluble in organic solvents, just like Resitol. It is normally ground to a powder and mixed with hexamine (hexamethylene tetramine) which provides further methylene linkages to act as cross-links and so converts the soluble and fusible Novolak to a hard, infusible solid.

There are many resins belonging to this class which are made by mixing compounds with hydroxyl groups attached to aromatic molecules with any of the many aldehydes.

The whole process, of course, can be carried out with hexamine only. In this case no catalysts are required. Hexamine ($N_4(CH_2)_6$) decomposes into formaldehyde and ammonia; further polymerization is promoted with methylene bridges derived from this formaldehyde. It has been established that a nitrogen-free resin is formed by heating 6 phenol:1 hexamine in molar ratio and that all the nitrogen in hexamine escapes as ammonia during polymerization. This indicates that hexamine alone is responsible for the formation of methylene bridges.

The mechanism of the hardening process using hexamine has been summarized by Zinke (1951). It was proved that the nitrogen content of the resin depends on the molar ratio of the reactants, polymerization temperature and time sustained during polymerization; resins free from nitrogen were formed only with excess phenol. According to Zinke, nitrogen in the resin exists in the form of dimethylene–amino bridges. In a resin with excess phenol, nitrogen is eliminated in ammonia. This results in the formation of methylene bridges giving a phenol–methylene chain shown in fig. 3.

The effect of heat-treatment depends on the degree of cross-linking. Thus, when weakly cross-linked material is heated slowly, low molecular weight substances, probably excess phenol and small-chain polymers, and water, are evolved between 100 °C and 350 °C. Above 500 °C small quantities of carbon monoxide and methane are evolved, but the final carbon yield is high : > 85%. If the material is highly cross-linked, water is not evolved until after 400 °C.

Fitzer *et al.* (1969) consider that between 100 and 300 °C polymerization continues to form long-chain polymers. Towards 300 °C hydroxyl groups on neighbouring chains condense to form an ether linkage. Simultaneously, the hydroxyl group may react with the methylene group in a neighbouring chain to form a closer cross-link akin to the diphenylmethane structure.

Water is formed from the free hydroxyl groups. Subsequently, the methylene bridges rupture, forming methane as a side product and the furan ring is opened. This is detected by the release of carbon monoxide, dioxide, and further water. The furan ring is stable up to 275 °C. Between 300 and 400 °C the C—O—C infra-red band is reduced in intensity due to ring rupture. Half of the furan ring oxygen is released as water and half as carbon monoxide or carbon dioxide. At 400 °C new infra-red bands appear, indicating the build-up of aromatic systems formed spontaneously from fragments of the furan ring. Above 450 °C a reaction between eliminated water and remaining methylene bridges occurs to form keto groups. Methane ceases to be formed as the methylene bridges are oxidized and significant amounts of hydrogen are released. Above 460 °C the carbonyl groups liberate carbon monoxide leading to a highly unsaturated aromatic residue capable of forming a cross-linked aromatic network.

It is difficult to follow the detailed variations of the chemical structure of the resin during pyrolysis from the infra-red studies alone; they can only lead to tentative interpretations of the chemical reactions taking place in a material in which various complicated chemical reactions must take place simultaneously. It is necessary to carry out chemical analyses, escaping gas analyses and identification of volatile products for the detailed understanding of chemistry of carbonization. Our interpretation is deduced from the scientific literature on the polymerization of phenol–hexamine and the carbonization of phenol–formaldehyde, particularly those carried out by Ouchi and his co-workers (1955, 1956, 1959, 1966) and from our own work on phenol–hexamine based resins (Kawamura and Jenkins, 1970).

It is also difficult to analyse the infra-red absorption spectra quantitatively because high background absorption appears at a heat-treatment temperature of 500 °C, due to $\pi \to \pi^*$ transitions, that is, transitions in which a π-electron is excited to an antibonding π-orbital. These occur in compounds which contain conventional double or triple bonds or aromatic rings. The energy required corresponds to wavelengths in the ultra-violet region. Increasing conjugation moves the absorption to longer wavelengths and finally into the visible region.

Fig. 4. Possible reactions taking place during the pyrolysis of phenolic resins.

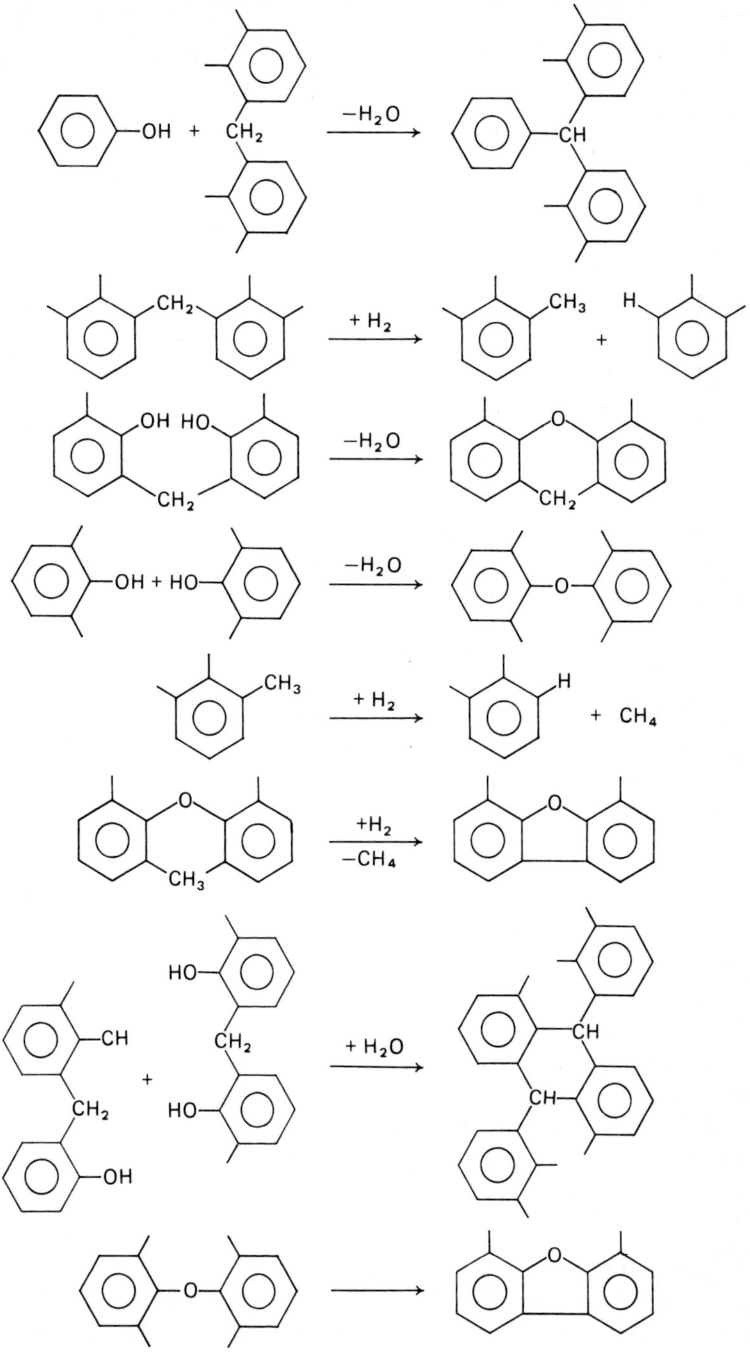
OH
CH2
−H2O
CH
CH2
+ H2
CH3
H
+
OH HO
−H2O
O
CH2
CH2
OH + HO
−H2O
O
CH3
+ H2
H
+ CH4
O
+H2
−CH4
O
CH3
HO
CH
CH2
+ H2O
CH
CH2
HO
CH
OH
O
O

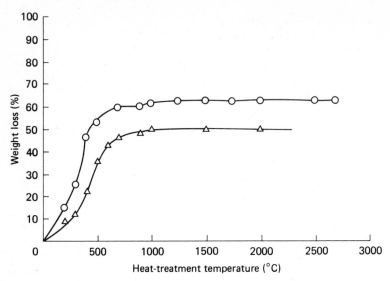

Fig. 5. Weight loss of phenolic resins during pyrolysis. ○, 12 phenol:1 hexamine; △, 6 phenol:1 hexamine.

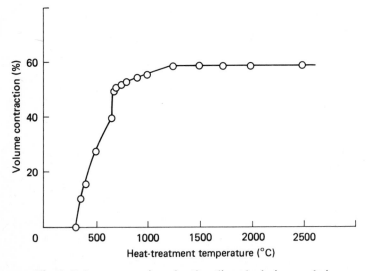

Fig. 6. Volume contraction of a phenolic resin during pyrolysis.

Despite these difficulties it is possible to show from the infra-red results that intermolecular cross-links are forged at 350 °C. Above this temperature the only changes are the progressive disappearance of aliphatic ether linkages ($-CH_2-O-CH_2-$) towards 500 °C, and a progressive increase in the degree of substitution in aromatic groups. The various possible reactions are illustrated in fig. 4.

Figure 5 shows the weight loss of a 12:1 resin as a function of heat-treatment temperature. A marked weight loss is observed between 300 and 400 °C which corresponds to the change in chemical structure detected by infra-red spectroscopy. The rate of weight loss decreases gradually up to 1250 °C and thereafter the weight loss remains constant at 62.5% of the weight before heating.

The weight loss of a resin is highly dependent on the molar ratio of phenol to hexamine. The weight loss of the 6:1 mixture follows the same shape as that of the 12:1 mixture but the final weight loss is reduced to 49.6%, because the 12:1 mixture contains excess phenol molecules which cannot be linked together with methylene bridges.

The variation of volume contraction with heat-treatment temperature is shown in fig. 6 for the case of phenolic resin discs. There are three features: a 40% contraction due to the removal of large molecules such as water below 500 °C; an abrupt stepwise contraction of 10% between 650 and 700 °C, and a steady contraction at higher temperatures, corresponding to the removal of hydrogen. The sudden collapse at 700 °C

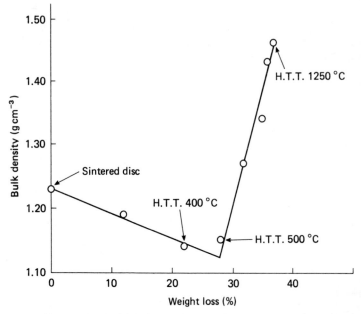

Fig. 7. Changes in bulk density in phenolic resin during pyrolysis plotted against weight loss.

is associated with a rapid rise in electrical conductivity. The accessible micropore volume also decreases rapidly above this temperature (Fitzer and Schäfer, 1970), due to the closing of pore entrances.

The carbonization and dehydrogenation stages are easily distinguishable by plotting the bulk density against the weight loss in sintered discs of phenolic resin: see fig. 7. Below 500 °C, a weight loss corresponds with a decrease in bulk density. Above 500 °C, the density increases rapidly with the removal of a relatively small mass of material – mainly hydrogen.

The kinetics of the thermal degradation of phenolic resins has been studied extensively. For instance, Friedman (1963), using TGA at linear rates of temperature rise, proposes a rate law for the carbonization stage:

$$\ln\left(-\frac{1}{w_0}\cdot\frac{dw}{dt}\right) = \ln A + 5\ln\frac{w - w_f}{w_0} - \frac{\Delta E}{RT}$$

where w = weight of organic material, w_0 = weight of original plastic, w_f = weight of final carbon, A = the pre-exponential factor ($=10^{18}\,h^{-1}$), ΔE = activation energy = 55 kcal.

This assumes a single simple thermally activated mechanism of constant activation energy. The high 'order of reaction' (5) serves merely to indicate the rapidity with which a weight loss characteristic of a given temperature is approached. In reality, there must be a sequence of reactions each with a characteristic pre-exponential factor and activation energy.

Work at Swansea on TGA (Easton and Jenkins, 1974) of the pyrolysis of phenolic resins at different rates of temperature rise (from 0.1 to 2.0 °C per min), showed that all curves are superimposable if we plot weight loss against temperature. This must be interpreted as indicating that at any given temperature the chemical process of pyrolysis moves to completion so rapidly that the chemical kinetics really have no effect on the weight-loss curves.

We have worked mainly with this material in the production of glassy carbon and fibre, and so the pyrolysis will be referred to continually in the following chapters by way of example for the whole range of polymeric carbons.

2.2.4   Polyfurfuryl alcohol (PFFA)

Polyfurfuryl alcohol is a thermosetting resin obtained from furfuryl alcohol (FFA, illustrated in fig. 3) which was formerly manufactured by the hydrogenation of furfural in the presence of copper oxide–chromium oxide catalyst. More recently, however, production has been facilitated by continuous operation in the vapour phase under very nominal hydrogen pressures.

Resinification takes place in the presence of acids, the reaction being exothermic and having a high temperature coefficient; adequate means for heat removal must be provided. The reaction may be stopped at any desired point by cooling and neutralization of the acid. Furfuryl alcohol catalysed by 3% maleic acid polymerizes exothermally at 110 °C. The reaction involves the alcohol group from a furfuryl alcohol molecule with an active hydrogen from the ring of an adjacent molecule forming difurfuryl alcohol. This is followed by the reaction of two adjacent alcohol groups to form difurfuryl ether, which, on splitting off formaldehyde, is converted to difurfuryl methane. The resulting formaldehyde is re-absorbed forming methylene and ether bridges between nuclei in adjacent chains.

Fitzer and Schäfer (1970) have examined the thermal degradation of such a polymer and found that the total weight loss depends on modifications to the alcohol, varying from a maximum of 54% for the pure alcohol. FFA–formaldehyde resins give smaller yields. The most rapid rate of weight loss occurs at 400 °C. The proposed mechanism of pyrolysis is shown in fig. 8.

The resin is used as a precursor for glassy carbon by Tokai Electrodes (BP 1033277, filed 1966) and General Electric (BP 921236, filed 1963). It is used extensively as an impregnant for densifying electrographite electrodes.

2.2.5 Polyacrylonitrile (PAN)

This is produced by addition polymerization of acrylonitrile and has the formula $(CH_2 \cdot CH(CN))_n$. Commercial PAN is atactic and so has two-dimensional order only, in a plane perpendicular to the fibre axis. There is no order along the fibre axis and so no true crystallization can take place.

Because of the low dye affinity of the pure polymer, the manufactured fibre usually consists of a copolymer with 10% vinyl acetate, vinyl chloride, styrene, isobutylene or acrylamide. The degree of polymerization (n) achieved in Orlon produced by du Pont, or Courtelle by Courtaulds is higher than 2000, which renders the material suitable for fibre manufacture. Dralon T manufactured by Farbenfabriken Bayer is a homopolymer containing 99 to 99.5% acrylonitrile. The main problem in producing filaments is the inability of the polymer to dissolve in

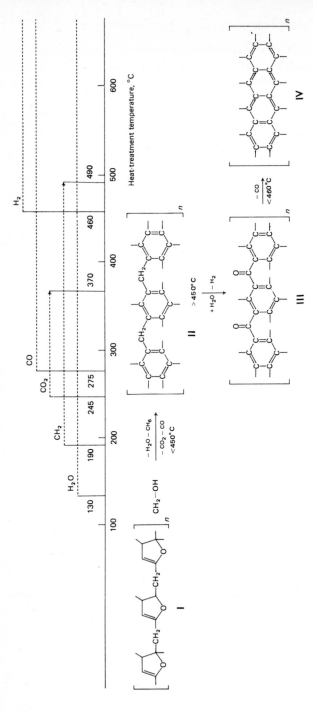

Fig. 8. Pyrolysis chemistry of polyfurfuryl alcohol (after Fitzer and Schäfer, 1970).

normal solvents suitable for solvent spinning. This is because of strong hydrogen bonding which increases the intermolecular cohesion so that despite the intervening —CN group, the carbon chains approach each other to a distance of 5.3 Å. Any solvent must therefore be strongly polar.

Moreton (1970) described how PAN fibre may be manufactured. He used 6% methyl acrylate as comonomer and spun the fibres from 8% by weight solution of PAN in 50% by weight aqueous sodium thiocyanate. Argon pressure was used to extrude the polymer solution through a single spinneret hole into a coagulating bath containing 10% aqueous thiocyanate. The speed of guide rollers was adjusted to stretch the length of the fibre by three as it passed through the coagulation bath. After washing at 50 °C, the fibre was hot-stretched as it passed through a steam tube. This increases the preferred orientation of the polymer chains.

In order to aid spinning, it is normal to coat the fibre with a finish or 'size'. This can be removed by continuous extraction with diethyl ether (Ezekiel, 1971), to yield a shiny white fibre.

When heated above 160 °C, PAN has the unique property of forming a copper-coloured ladder polymer consisting of fused hydrogenated naphthyl–pyridine rings with the formation of a conjugated —C—N—C— chain, as illustrated in fig. 9. The reaction is exothermic, producing sufficient heat at 280 °C, for instance, to cause complete fusion of the polymer and complete loss of orientation of polymer chains. The extent to which the length of conjugated chain is formed depends on the tacticity. The formation of the ladder chain enhances the thermal stability since to get small volatile fragments four bonds must be broken in neighbouring rings.

For some time it has been known that PAN fibres blacken on heating in air at 220 °C and that this 'black Orlon' is remarkably flame-proof and stable. Such stabilization by oxidation permits the formation of oriented ladder polymer whilst reducing the intensity of the exothermic peak, which would otherwise build up if the oxidation process were to be omitted. After stabilization, the modified ladder molecules have a high enough glass transition temperature to retain any preferred orientation imposed in the original polymer, even subsequent to a carbonization process (see Watt, 1972).

Watt (1970) suggests that some of the methylene groups (—CH$_2$—) in the original polymer are converted to ketonic groups (—CO—) by oxidation to produce the modified ladder polymer illustrated in fig. 9.

As to the subsequent carbonization process, the weight of pre-oxidized homopolymer fibre decreases to 53.6% of the original value, most of the

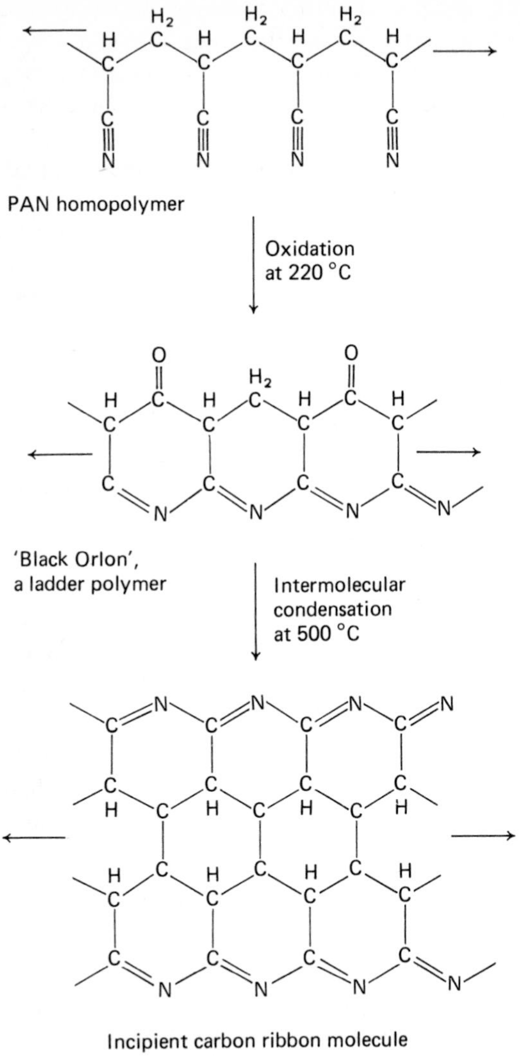

PAN homopolymer

Oxidation
at 220 °C

'Black Orlon',
a ladder polymer

Intermolecular
condensation
at 500 °C

Incipient carbon ribbon molecule

Fig. 9. Possible intermediate structures present during pre-oxidation and pyrolysis of polyacrylonitrile.

decrease occurring below 500 °C with evolution of H_2O and CO_2. In contrast to phenolic resins, there is a simultaneous rise in specific gravity from 1.2 to 1.6. Above 500 °C, the density increases progressively to about 1.95 at a heat-treatment temperature of 2500 °C.

Stiffness increases progressively with elimination of hydrogen above

500 °C. Nitrogen is also eliminated at this stage but this process is not progressive, there being a sharp peak rate of nitrogen evolution at 900 °C which is not reflected in a stepwise increase in stiffness.

The proposed mechanism of carbonization is illustrated also in fig. 9. The ladder molecules coalesce progressively to form the ribbons characteristic of all polymeric carbons. It should be stressed that the structures drawn in fig. 9 are essentially related to isotactic homopolymers, whereas the fibres used to make high stiffness, high tenacity carbon fibres are atactic copolymers. This makes the true mechanism difficult to interpret.

The commercial process for making carbon fibres from PAN is extremely important and is outlined in chapter 3.

2.2.6 Hydrocarbons

The pyrolysis of hydrocarbons has been adequately described by Fitzer *et al.* (1969*a*). We need only underline here the matter relevant to our subject.

There are two forms of hydrocarbon occurring naturally: paraffins, consisting of flexible chains of sp_3 carbon atoms, and aromatics, consisting of stiff rings of sp_2 carbon atoms. On heating paraffins, a series of reactions occur in sequence: thermal degradation into smaller active units, dehydrogenation, cyclization and aromatization to form stable aromatics. The aromatics undergo chemical condensation to form polycyclic aromatics which crystallize to form pre-cokes.

In the commercial process known as 'delayed' coking, petroleum hydrocarbon fractions are heated rapidly to about 430 °C under pressure and maintained at this temperature and pressure for some time, perhaps over a few days. Gradually, liquid crystals of polycyclic aromatics are formed and grow. Lowering the pressure and temperature afterwards allows the mesophase to be removed as 'green' coke. In this state the material still contains loose hydrocarbon molecules which can be removed with suitable solvents. It can even be compacted under pressure to produce strong compacts. 'Calcining' to 1200 °C produces true coke from which all residual hydrogen has been removed.

The contrast between the aromatizing process in such hydrocarbons and that in polymeric systems should be underlined. The coking process is slow and can take many days to complete. The charring process is rapid and can achieve a set stage characteristic of a given temperature within minutes of reaching that temperature.

Additions of sulphur to and oxidation of petroleum pitches, by air blowing, for instance, increase carbon yield but lower the crystallinity of the final coke. Large additions of sulphur will produce lower density non-graphitizing material similar in every way to dense isotropic polymeric chars. It is presumed that polymeric material is produced in the precursor by the formation of oxygen or sulphur linkages.

Pyrolysis of pure aromatic hydrocarbons has been extensively investigated (cf. Edstrom and Lewis, 1969). Naphthalene, anthracene, isotruxene and acenaphthalene all form highly crystalline graphitic cokes. However, another group of aromatics, typified by phenanthrene, biphenyl and truxene, form non-graphitizable chars similar to those derived from polymeric precursors. It is tempting to postulate two modes of coalescence or condensation to form polycyclic aromatics. The one is co-planar with circular symmetry and no preferred direction of growth, to form large easily-packed polycyclic rafts; the other is also co-planar but with a preferred axis of growth, forming easily-bent, warped and interwoven polycyclic ribbons which cannot be unravelled subsequently to form the extensive sheets necessary for the production of graphitic material.

Fibres from petroleum pitch fractions can be made by melt extrusion and stabilized by oxidation prior to carbonization to form cheap carbon fibres (Kureha Chemical Co. of Japan).

Coals are another natural source of hydrocarbons which are almost completely aromatized and associated with some oxygen. They are derived by prolonged heat and pressure from plant materials such as cellulose and lignin but there is still disagreement about the coalification process.

Coals are classified with respect to the 'rank' or stage which they have reached in this natural process of coalification which is really a long delayed condensation process under pressure to produce larger and yet larger polycyclics. In terms of volatile content, 'high volatile' coals have more than 30% volatile matter, 'medium volatile' have 20–30%, 'coking steam' coals have 14–20%, 'dry steam' coals have 10–14% and 'anthracites' have less than 10% volatile matter. In terms of carbon content, 'lignites' have less than 84% carbon, bituminous coals have 85–91% carbon and anthracites have more than 91% carbon.

Coals of lowest rank (lignites) are said to possess a polymeric character in that aromatic lamellae are linked by bridges. As coalification advances these bonds are supposed to gradually disappear while aromatic lamellae start growing in size. In the range 85–91% carbon the aromatic lamellae possess the highest degree of mobility at coking temperatures. Above 91% carbon the crystallites have such a high thermal stability that the coal does not soften when pyrolysed.

On pyrolysis, coals of low rank form isotropic chars (polymeric carbons), bituminous coals produce finely textured crystalline cokes, while anthracites, although not coking in the normal sense, eventually yield graphites at sufficiently high temperatures.

Untreated coals have a high ash content and these impurities make raw coal unsuitable as precursors for carbons. Work is progressing into solvent extraction which results in a pure hydrocarbon extract from which both highly crystalline cokes and carbon fibre can be derived.

3 Fabrication

3.0 Production of glassy carbon ware: generalities

When polymers are carbonized without taking special precautions, the end result is either a misshapen porous mass of carbon expanded and distorted by the gaseous products which are an inevitable result of carbonization, or a pile of sharp fragments – debris of the effect of rapid gasification. Once this useless mass is formed, it is impossible to reform it into a desired object because carbon does not exhibit any plastic deformation and therefore cannot be melted or sintered at temperatures below 2700 °C. The shaping process must take place prior to carbonization. Naturally, in removing elements other than carbon there will be a large but predictable contraction which must be accepted. The subsequent heat-treatment must allow gases to escape without distorting the original shape.

The oldest way is to carbonize a phenolic resin. The 'char', so formed, is ground to suitable dimensions and hot-mixed with a resin binder of, say, partially polymerized phenol–benzaldehyde. The hot mix of filler and binder is extruded or pressed to the desired shape and fired thereafter to 1000 °C in a calcining furnace to produce so-called 'hard' filler, 'hard binder' carbons. This is identical in the early stages to the production of electrographite blocks. Care is taken in the relationship between filler particle size and overall dimensions to allow gases to escape. For large blocks, a large particle size is essential. For small blocks, finer particles are possible. The heating programme must also be carefully controlled to ensure that gaseous products do not build up sufficiently high pressures in internal gas pockets to disrupt the blocks, allowing time for the gases to diffuse to the surface. The optimum density can, of course, never be achieved in one such operation. The blocks are densified by impregnation with suitable impregnants such as furfuryl alcohol and are then refired. Even so, the permeability remains high, and the strength is inferior to the more recently developed monolithic carbon glasses (Lewis *et al.*, 1963).

Davidson (1962) of The General Electric Company succeeded in making tubes wholly out of glassy carbon from cellulose. Yamada and Sato (1962) produced similar material from phenolic resin. Since then, other applications have been found, and now artefacts from this material are being manufactured by many companies using a variety of patented and/or secret processes. The following is a list of patents which typify the many processes which have been found successful:

Davidson, H. W. (GEC Ltd)   *filed* 1961
   BP 860342
Davidson, H. W. (GEC Ltd)   1962
   BP 889351
Rivington, H. L. (GEC Ltd)   1963
   BP 921236
Redfern, B. (Plessey)   1964
   BP 956452
Redfern, B. (Plessey)   1966
   BP 1024971
Societé le Carbone–Lorraine   1966
   BP 1031126
Yamada, S. (Tokai)   1966
   BP 1033277
Jenkins, G. M.   1972
   BP 1228910

Patents are registered by Tokai Electrodes (Yamada, 1968) to cover the processes involved in making artefacts out of glassy carbons, depending on the size and shape of the desired product. They also found it possible to make small hollow spheres by blowing the condensates through a cloth prior to hardening and carbonizing. The diameter of the bubbles thus obtained can vary between 0.05 and 1.0 mm. Surface tension during gelation ensures that they are almost perfectly spherical.

It is reported that 'Vitro Carbon' artefacts are made at the Resources Research Institute, Japan, by resinifying and hardening a mixture of acetone and furfural in molar ratio 2:1, which is aged for a few months before carbonization (Honda and Samada, 1966).

Bradshaw and Pinoli (1967) of Lockheed, USA, reported that their glassy carbons are made from naphthalenediol using pressures during carbonization of between 100 and 700 atmospheres over prolonged periods. The patent option of L. Rivington (GEC) (BP 1024971, filed 1963) also proposed baking under pressure of 140 atmospheres at temperatures up to 400 °C for three days prior to carbonization, starting with furan, cumarone, indene or cyclopentadiene.

During carbonization, a relatively large weight and volume change occurs and it is the manner in which this happens which affects the quality of the product. Cracks, blisters and distortions often appear in the material and to minimize these defects it is necessary to determine the major causes. These may be any combination of the following: thermal expansion of the artefact; contraction or shrinkage of the artefact; passage of pyrolysis products through the artefact. These effects can all be reduced to acceptable levels by using sufficiently slow rates of heating, and although such a solution is to be avoided as far as possible, for economic reasons related to throughput, it is, nevertheless, the only solution so far available.

In the patents concerning the cellulose-carbon process no exact rates of heating are specified. It is, however, stated that a period of perhaps as long as ten days is necessary in some cases for the initial carbonization process. If this is averaged out, it amounts to approximately 4 °C per hour. It is understood from this patent that the main cause of disruption, when it occurs, is the passage of pyrolysis products through the artefact.

Since the volume of gases permitted to travel in a given time through a microporous medium is proportional to the pressure drop across the structure, it is argued that an increase in the outside pressure should prevent a disastrously rapid outflow of decomposition products. A pressurized process would enable the treatment period to be considerably reduced. This has been attempted by Davidson (1962) and reported to be the case. The artefact is loaded into a pressure vessel which is purged of air and filled with nitrogen at 100 atmospheres. The vessel is then heated to 360 °C over a period of 48 hours, the pressure rising to about 200 atmospheres. The pressure is reduced at this temperature and, when normal atmospheric pressure is reached, the vessel is cooled down and unloaded. The temperature of 360 °C is a limitation of the vessel since it may be overstrained at higher temperatures at this pressure. Nevertheless, the majority of evolution products are given off below this temperature and it was considered by Davidson that this type of processing was a marked improvement.

In the GEC patent (BP 92123, filed 1963) for carbon from furfuryl alcohol, the resin (which has been cured at 400 °C under nitrogen pressure) is heated to 1000 °C at a rate of 300 °C per hour *in vacuo* or in nitrogen. With respect to this patent, the initial curing stage is the more important and it can be inferred from the high heating rates that the proportion of decomposition products after this curing will not be high.

In the patent issued to the Plessey Company (BP 956452, filed 1964) for the production of a carbon from a phenolic resin, a heating rate of 5 °C per hour is quoted most frequently, although in one of their examples a heating rate of 1 °C per hour is given. Above 600 °C the rates are increased progressively with temperature, the final temperature of 1400 °C being maintained for 24 hours. Pressure carbonization is not considered.

A recent paper by Fitzer *et al.* (1969) casts doubt upon the usefulness of pressurization during carbonization, and results quoted indicate that pressure in the region of 20×10^6 atmospheres would be required to slow down the transport of the released gases appreciably. This suggests that the diffusion problem is not that of diffusion through a microporous structure, where mean pressure and pressure gradient are the dominating factors, but may be a problem in transmission of fluid through a polymeric barrier. This is known as '*selective solubility diffusion*' and the rate controlling mechanisms are considered to consist of: adsorption and solution of a component into the polymeric matrix; diffusion of solute through the polymer under a concentration gradient; desorption and evaporation of solute from the surface of the polymer. The polymer may possess a certain amount of porosity but the flow through these closed pores will not, in general, be a major portion of the total flow.

The question of the controlling diffusion mechanism is worthy of further investigation, since the only other explanation is that during the intermediate stage, between pure polymer and pure carbon, the material exhibits open pores. It is postulated that these open pores then close.

In the author's laboratory, A. Easton has shown that the reaction kinetics of carbonization of phenolic resin rods, for instance, is heat transfer controlled (Easton and Jenkins, 1974). Results indicate that the 'thermal reforming' process is predominant between 300 and 500 °C and is endothermic. It is necessary to supply heat and so the rate of heat transfer is the rate controlling factor. This leads to an 'unreacted core' which is found even at low rates of temperature rise.

Easton has also shown that thick samples of glassy carbon cannot be made because of fissuring which takes place at temperatures above 280 °C in the case of solid cylindrical rods of phenolic resin – in the region where one observes the greatest evolution of gaseous products. The temperature at which fracture occurs (T) varies inversely as the diameter of the rod (D) for a given rate of temperature rise ($d\theta/dt$), such that

$$T - T_c = C/D - D_c$$

where $T_c = 280$ °C and $D = 5$ mm. C is found to vary as $\sim(d\theta/dt)^{-2/3}$.

This can be explained by attributing the disruption of the cylinders to the build-up of pressure from gaseous products which need time to percolate through the outer layers. The greater the diameter, the greater the time needed for gases to diffuse through. The faster the rate of temperature rise, the less time allowed for the gases to escape. Figure 10 illustrates how the temperature of fracture of cylindrical phenolic resin rods varies with cylinder diameter and heating rate during pyrolysis.

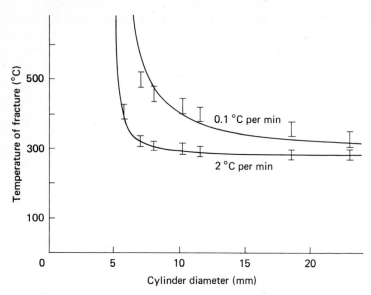

Fig. 10. Variation of the temperature of fracture of cylindrical phenolic resin rods varies with cylinder diameter and heating rate during pyrolysis.

3.1   Specific processes

3.1.1   GEC Process (UK) (Davidson and Losty, 1963)

Natural cellulose is available only in a fibrous form and this must be used to make a structure which will be preserved throughout the heat-treatment. The fibres must be packed together to form a dense solid to produce a carbon of reasonable density.

The fibres are separated and reduced in length to permit maximum packing density by prolonged mechanical beating under water in a conventional 'Hollander' machine. The heavily beaten fibres absorb twenty times their own weight of water and so only dilute suspensions can be beaten, 2% by weight being typical.

In order to remove sufficient free water to permit handling of the resultant solid and to lay down the fibres to produce optimum packing, a tube of packed fibres is slip-cast in the bowl of a conventional refining centrifuge. The pulp is pumped into the imperforate bowl while it is spinning so that the solids are thrown to the wall; clear water is carried over the top of the bowl. The cellulose fibres deposit as a thick even layer on the walls of the bowl and contain 90% water. The resulting tube is firm enough to be extracted, assisted by lining the bowl with a PTFE sleeve. The remaining water is removed in an air oven. Mandrels are used to preserve the geometry because of large shrinkages. When all free water is gone, the tube is machined to the required shape prior to carbonization.

Slip-casting using a centrifuge restricts the forms in which cellulose carbon can be fabricated and alternative methods are adopted to allow a variety of shapes of various thicknesses to be produced. The dense cellulosic bodies so obtained are impermeable to helium and to the gaseous products of decomposition in the early stages. Thus the body is heated extremely slowly to prevent these volatiles blistering the faces. If an external gas pressure is applied to balance the internal pressure, blistering will not occur and more rapid heating rates are possible. Once the material has developed appreciable porosity the pressure can be reduced.

To achieve these conditions the specimens are loaded into a cylindrical pressure vessel and pressurized under nitrogen. The system is sealed and kept under pressure during the heating cycle. During heating the pressure rises owing to the rise in temperature and the release of volatile products. When the material is sufficiently porous the pressure is released and the samples transferred to conventional furnaces. During carbonization under pressure, uniform shrinkages of over 25 % occur, with some distortion in thin-walled articles.

Such tubes are engaged tightly on to steel mandrels to correct this distortion, making certain that at the engagement temperature the cellulose is still shrinking. The cellulose must be plastic enough to conform to the shape of the mandrel. These conditions are achieved between 550 and 800 °C. Many articles are sufficiently rigid for heat-treatment up to 1500 °C in a single furnace.

3.1.2   Swansea process for simple glassy carbon artefacts

A technique has been developed in Swansea (Kawamura and Jenkins, 1970) for making simple artefacts out of phenolic resins. Basically, cross-linking between polymer chains is prevented until a late stage by using excess phenol in a phenol–hexamine system. The uncross-linked polymer is heated to temperatures a little below the temperature of onset of true carbonization. Excess phenol and unwanted material are thus eliminated. It is possible in this way to produce a clear polymer which can be shaped as desired at temperatures in the region of 300 °C. This can be heated quite rapidly ($\sim 3°\,min^{-1}$) to complete its carbonization at 1000 °C. The porosity can be adjusted by grinding the polymer into a powder at room temperature and sintering at 330 °C to give a desired porosity prior to carbonization.

We shall enter into more detail in this process because of our intimate knowledge of the technology involved and because it serves to illustrate the difficulties to be encountered in a typical glassy carbon forming technique.

A 12:1 phenol–hexamine does not maintain its shape during poly-merization and carbonization without special precautions; a typical heating rate of 0.1 °C min^{-1} is not sufficiently slow to prevent escaping gases destroying the original shape. The resulting carbons are useless as engineering materials and are also unsuitable for the measurement of variation of physical properties during carbonization.

It was observed that the major change of volume took place between 170 and 250 °C, even though a change in the chemical structure was not observed in infra-red spectroscopy studies up to 300 °C. It was concluded that resin pre-heated at about 300 °C in nitrogen could maintain a given shape during carbonization. It was decided to produce disc-shaped samples from pre-heated resins by a process based on powder technology. Resin pre-treated to 300 °C was crushed into fine powder, compacted to a disc shape and carbonized under the conditions described above. The resulting carbon disc showed slightly uneven expansion in the centre of the disc along the compacting direction; it was porous and the pore sizes were large.

Carbon discs prepared from resin pre-heated at between 300 and 320 °C showed a similar macroscopic structure. However, a carbon disc made from the resin pre-heated at 330 °C could maintain the disc shape almost perfectly during carbonization. The disc was still porous but the pore sizes were much smaller than for the other pre-heat-treatments. It was found that the pore size could be controlled using various particle sizes of the treated resin and various compacting loads. Nevertheless, the pores were not completely removed.

It was found that compacted discs made of resin pre-heated between 300 and 330 °C could be sintered under a small load. In this way the disc was rendered optically transparent. Such a disc could maintain its shape during carbonization without constraint and the resultant carbon disc was non-porous on a macroscopic scale.

The preparation steps for the production of porous and non-porous carbon discs are summarized below:

Pre-heating of the original resin at temperatures between 300 and 330 °C in flowing nitrogen gas.
Crushing the pre-heated resin into fine particles.
Compaction of the particles into a disc at room temperature; a compacting stress of 350 MN m^{-2} is generally employed.

Sintering of the compacted disc under a compressive stress of about $10^5 \, \mathrm{N\,m^{-2}}$ in flowing nitrogen gas from room temperature to the pre-heating temperature.

Carbonization of the sintered disc in flowing nitrogen gas; the heating rate is 1 to 5 °C min^{-1} up to 800 °C and 5 to 10 °C min^{-1} thereafter.

This process is applicable to the production of carbon products with other shapes and is patented by the National Coal Board (BP 1228910, filed 1968).

The particle size of the pre-heated resin is not critical for the production of the non-porous carbon discs, although it influences the physical properties of the porous carbon discs. In general, resin pre-heated at a lower temperature produces carbon discs with higher bulk density, provided that the same stress is applied during sintering. The bulk density of discs heated above 1000 °C can be varied between 1.35 g cm^{-3} and 1.50 g cm^{-3} according to the pre-heating temperature. The carbon discs with bulk density less than 1.4 g cm^{-3} have some porosity but the pores are not necessarily interconnected.

The pre-heating temperature must not exceed 350 °C because resin particles will then not compact even under very high stress. This indicates a marked change in the mechanical properties of the material which should correlate with a profound change in chemical structure involving the formation of intermolecular cross-links.

The heating rate during carbonization of the discs must be carefully controlled and depends on the pre-heating temperature of the original resin. Discs with bulk density less than 1.4 g cm^{-3} can be heat-treated at a heating rate of 5 °C min^{-1} throughout the carbonization process. A disc with higher bulk density needs a slower heating rate.

If the sintered polymer discs are carbonized under a compressive stress, the resulting carbon discs fracture on a macroscopic scale because diameter shrinkage of the disc is constrained.

Figure 11 illustrates the diameter and height shrinkage of a sintered polymer disc as a function of heat-treatment temperature. There is a rapid decrease in dimensions up to 700 °C, thereafter decreasing gradually to reach constant values at 1250 °C. The largest shrinkage rate is observed between 650 and 675 °C. Since X-ray diffraction studies do not detect any significant structural change at these temperatures it is considered that the large shrinkage is due to a chemical process such as dehydrogenation. Infra-red spectroscopy studies confirm this.

Both diameter and height show similar behaviour but the final diameter shrinkage is about twice that of height. If the shrinkage is due only

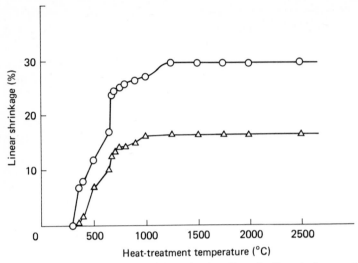

Fig. 11. Dimensional shrinkage in phenolic resin discs during pyrolysis. ○, diameter shrinkage; △, height shrinkage.

to chemical process such as cross-link formation, both diameter and height should follow each other in an isotropic material. The differential shrinkage must therefore be due to the internal strains imposed when the disc is compacted and sintered under a compressive stress.

A large differential shrinkage is observed between 300 and 350 °C. Infra-red spectroscopy studies show that no significant chemical changes take place below 350 °C. Therefore, the differential shrinkage must be due to differential shrinkage of the chain polymers which are probably strained along the cylinder axis during compaction and sintering, the work done being stored as internal strain energy.

3.2   Production of carbon fibres

3.2.1   Isotropic carbon fibres

Carbon fibre was first made commercially by Edison in the last century for filaments of electric light bulbs. He carbonized thin bamboo shoots and cotton, all consisting mainly of cellulose. The fibres were isotropic, and thus had low stiffness.

From early days it was known that forms of cellulose fibre, especially rayon fibre, polyacrylonitrile and Saran fibres, all carbonize to give carbon fibres.

Otani (1965) experimented with the production of isotropic carbon fibres from molten pyrolysis products such as polyvinyl chloride pitch and coal tar pitch. He was the first to show that it is possible to produce carbon fibres from various kinds of apparently non-polymeric starting materials; it is not necessary to choose textile fibres.

Polyvinyl chloride produces a molten pyrolysis product after being carbonized in nitrogen at temperatures between 390 and 415 °C. The pyrolysis product (PVC pitch) softens in nitrogen gas at temperatures above 150 °C and forms a viscous liquid at temperatures above 200 °C. The PVC pitch can be extruded as fibre using a simple melt extrusion process and the extruded fibres maintain their shape during carbonization if the surface of the fibres is sufficiently oxidized before carbonization. The resultant carbon fibres show no preferred orientation and a poor degree of graphitization.

Otani (1966) later found that ozone was very effective for the oxidation of the surface of the fibres; air required a higher temperature to produce sufficient oxidation of the surface of the fibres to preserve the fibre shape. He suggested that oxidation resulted in intermolecular cross-links because the surface of the fibre so treated is no longer soluble in chloroform. However, insolubility of the material in organic solvents does not always indicate the formation of intermolecular cross-links.

Both the tensile strength and the Young's modulus of PVC pitch based carbon fibres increase with decrease of the fibre diameter. The maximum values are $1.8\ \mathrm{GN\,m^{-2}}$ for tensile strength and $56\ \mathrm{GN\,m^{-2}}$ for Young's modulus. The Young's modulus increases gradually with decrease in diameter, while the tensile strength increases rapidly with decrease in diameter, especially below 15 μm; Otani gave no explanation for this.

Fibres derived from other pitches – petroleum and coal tar – are now made commercially by Kureha in Japan. The process probably closely follows the technique first worked out by Otani for PVC pitch, involving an essential oxidation step prior to carbonization.

Yamada (1967) first succeeded in making glassy carbon filament from a complex phenolic resin. Previously, no effort had been made to spin filaments from phenolic resins, which have low molecular weights. If the molecular weight is increased to over 10000, even these resins can be produced in filament form. It was found that a ternary system of Resol–Novolak–Furan resin had to be adopted to produce optimum spinning conditions. A typically successful system contained seven parts of a Resol derived from phenol–formaldehyde in molar ratio 1:2, catalysed by ammonia, one part of a Novolak derived from phenol–formaldehyde in molar ratio 1:0.88 and catalysed by hydrochloric acid, and ten parts of a furfuryl alcohol–furfural condensate in weight ratio 8:2.

The solution is heated and maintained at 70 °C. It is then spun under a pressure of $0.1\ MNm^{-2}$ of nitrogen through a nozzle diameter of between 0.3 and 1.0 mm. The extruded filament is cooled to 5 °C on the winding spool immersed in 10% HCl and is maintained thus for a few weeks, to be followed by washing, drying, and final carbonization. The size and shape of the nozzle affect the homogeneity and diameter of the filaments.

The fibre was found to be oval in cross-section. Improved die shape eliminated this effect and narrowed the variation in filament diameter, allowing even thinner filaments to be made.

Both strength and modulus could be improved by reducing the diameter and increasing the curing time from 4 to 12 days. Treatment with both sulphur and ozone led to a doubling of the strength, and so did carbonization of the filament stretched between two fixed points, but no increase in modulus was observed.

A simple melt extrusion process was employed by us for the preparation of phenol–hexamine fibres (Kawamura and Jenkins, 1970). The original resin (12:1) was heated in a stainless steel pot to temperatures controlled between 120 and 135 °C under nitrogen. The nitrogen gas pressure was increased so that the polymer was forced through an orifice at the bottom of the pot. The fibre was drawn in air by wrapping it around a rotating frame. It was found that the speed of rotation of the frame and the diameter of the orifice governed the diameter of the extruded fibres. Provided that the extrusion temperature was well controlled, the extrusion of the resin was simple and the extruded fibres were easily removed from the bobbin after cooling in air at room temperature.

In this way fibres of very large length could be drawn easily, with a minimum diameter of less than 20 μm. It was also possible to produce fibres with large diameters of a few hundred μm by this process. Carbonization of the extruded fibres was difficult because, during carbonization, the organic fibres tended to stick to each other and also to a metal tray on which they were placed. In addition, the heating rate during carbonization had to be carefully controlled in order to maintain the fibrous shape of the sample. The extruded fibres were placed separately on a porous asbestos tray and were carbonized under flowing nitrogen gas with a heating rate of 1 °Cmin^{-1} up to 400 °C, whereafter the fibres ceased to adhere; a bundle of them could be further carbonized without special precautions with a heating rate of 5 °Cmin^{-1} up to any temperature.

The length of the carbonized fibres was limited because the extruded fibres contained excess phenol, resulting in inhomogeneous polymerization of the material. In order to avoid this, a phenol–hexamine mixture without excess phenol was also tried. A 6:1 mixture in molar ratio was polymerized in the same way but it was found that control of viscosity was very difficult during polymerization. It was necessary to employ a more sophisticated technique for controlling the viscosity and preparing a resin suitable for extrusion.

It was possible to harden or cure the surface of the extruded fibres using a suitable catalyst before carbonization; separation of individual fibres was then unnecessary during the early stage of carbonization and production of carbon fibres of great length was possible.

Long lengths of carbon fibre with large uniform diameters of about 40 μm were easily produced by this method. Preparation of thin carbon fibre, of diameter less than 10 μm, was difficult, and the length limited. All the fibres had perfectly circular cross-sections.

Using this simple carbonization process, it was possible to produce carbon fibres from phenol–hexamine on a laboratory scale without any special surface treatment before carbonization and this was patented by the National Coal Board (BP 1228910, filed 1968). Although much technological development is necessary for the practical application of such fibre, important information was obtained on the formation of glassy carbon from the polymer by studying changes in the structure and physical properties of the fibre carbonized at various temperatures.

Adams *et al.* (1970) produced carbon fibres from commercial Saran (PVDC) fibres. These fibres are porous and, therefore, the tensile strength is low (0.5 GNm^{-2}); the Young's modulus is also low (30 GNm^{-2}). These mechanical properties might be improved if commercial Saran fibres with thinner diameters were available; 65 μm is the minimum diameter of commercial Saran fibres.

3.2.2   High modulus fibres

The basic principle in producing a high axial modulus in carbon fibres is to align the carbon–carbon bonds parallel to the fibre axis. The ultimate stiffness is, of course, that of graphite whiskers which have perfect alignment of carbon atoms and an axial Young's modulus of 1000 $GN m^{-2}$ – the highest in nature.

This alignment of carbon atoms can be achieved by plastically stretching polymeric carbons – either at high temperatures equivalent to the graphitization temperatures of graphitic carbons, or by the alignment of carbon chains in the precursor prior to carbonization.

High temperature stretching. Using fibres manufactured from a rayon precursor, Bacon and his co-workers (1965) showed that fibre 'crystallites' could be oriented parallel to the fibre axis simply by pulling at temperatures above 1800 °C. Later (1967), Bacon showed that on

TABLE 2   *Properties of commercial carbon fibres derived from Rayon*

Fibres	Manufacturer	Diameter (μm)	Density (gcm^{-3})	Tensile modulus (GNm^{-2})	Tensile strength (MNm^{-2})
Thornel 25	Union Carbide	7	1.42	160	1260
Thornel 50	Union Carbide	6.5	1.67	340	2030
HMG-50	Hitco	6.0	1.72	350	2170
Thornel 50	Union Carbide	5.6	1.82	550	2450

extending carbon fibres derived from rayon by 300% at 2750 °C under loads of only 1 g per fibre, the tensile modulus increased to 630 GNm^{-2}. The tensile strength was greater than 3.5 GNm^{-2}, which is the highest value ever recorded.

This is the basis for the commercial production of carbon fibre by Union Carbide, Polycarbon, Carborundum and Hitco in the United States. The fibres, characteristically, have crenulated irregular cross-sections and show increases in strength in step with increases in axial modulus. The density is surprisingly low. The properties of commercial fibres are outlined in table 2. Alignment of carbon atoms increases the tensile modulus by an order of magnitude but the strength rises only by a factor of two.

Linger *et al.* (1970) induced preferred orientation by hot-stretching isotropic carbon fibres derived from melt extruded pitch. The resultant carbon fibres had tensile strengths of up to 3000 MNm^{-2} and a tensile modulus of up to 500 GNm^{-2}. It may be inferred that any polymeric carbon fibre is plastic above 1800 °C and high preferred orientation can be induced in all such material by stretching to extension ratios of 3 or 4.

Stretching prior to carbonization. Preferred orientation of polymer chains can be imposed easily in many polymeric systems, especially polymer fibres, merely by stretching. The problem is to fix this preferred orientation so that it remains after carbonization. This has been achieved so far only in one polymeric system – polyacrylonitrile. Preferred orientation, as indicated in section 2.2.5, can be indelibly impressed in PAN fibres merely by oxidizing the stretched fibre at 220 °C.

Good results were first obtained by Shindo (1964) who found that PAN yielded a material with high Young's modulus. In the United Kingdon, two groups – Watt *et al.* (1966) and Standage and Prescott (1966) – made the discovery that if PAN fibres are stretched in air prior to carbonization, the resultant fibres possessed a remarkably high Young's modulus.

TABLE 3   *Properties of carbon fibres derived from PAN fibre oxidized under varying loads*

Loading applied to yarn (g)	Percentage length change at 220 °C	Axial tensile strength (MNm^{-2})		Axial modulus (GNm^{-2})	
		Heat-treatment temperature			
		1000 °C	2500 °C	1000 °C	2500 °C
Nil	−40	700	560	91	210
10	−12	700	700	112	266
20	+ 2	840	840	140	329
30	+15	1400	1400	147	371
40	+36	1400	1400	147	420

Early workers on this system (Shindo, 1961) thought that homo-polymer fibres of high molecular weight and high tenacity, higher than that of normal commercial fibres, would be most suitable for the production of carbon fibres with high preferred orientation. Watt (BP 1 110 791, filed 1968) showed that fibre similar to ordinary industrial Courtelle copolymer fibre gave better results because it could be stretched substantially during the oxidation stage or in the presence of steam prior to oxidation.

The original patent stated that 2.5 denier* PAN fibre, in the form of yarn of 100 filaments, was heated in air for 24 hours at 220 °C with varying loads applied. Complete oxidation was important; insufficient oxidation gave fibres of insignificant strength. The yarn was then carbonized in a non-oxidizing atmosphere up to 1000 °C and subse-quently heat-treated to 2500 °C.

Fibres oxidized free of stress shrank by as much as 40% in length. The carbon fibres ensuing had reasonable strength but low stiffness. If, however, a tension is applied during the oxidation process, the com-bined effect of oxidation and tension results in carbon fibres of remark-able stiffness, as indicated in table 3.

Because fibres show a remarkable shrinkage during oxidation, con-siderable tension can be imposed progressively on the fibres merely by winding them on a former prior to the oxidation process, thus preventing the shrinkage that would otherwise occur. Watt found that 100 filament 2.5 denier PAN fibre yarn, wound on a carbon former under slight initial tension and subsequently oxidized for 22 hours at 220 °C, gave excellent carbon fibre after carbonizing and annealing free of stress at temperatures up to 2700 °C.

* Denier is the weight in g of 9000 m of fibre. Knowing the density, it is a measure of the average cross-sectional area of a fibre.

Fibres thus formed and heat-treated to all temperatures above 1000 °C had tensile strengths of about 1800 MN m^{-2}. The axial Young's modulus increased from 140 GN m^{-2} at 1000 °C to 350 GN m^{-2} at 2500 °C and 420 GN m^{-2} at 2900 °C heat-treatment temperature.

The work of Watt at the Royal Aircraft Establishment, Farnborough, is the basis of the technology developed by Morganite–Whittaker, Courtaulds–Hercules, and Rolls–Royce for the commercial production of carbon fibre. Characteristically, the fibres have round cross-sections with some fluting. Similar fibres are made by Tokai and Toray in Japan.

It is a curious feature of these fibres that although strength and modulus increase together up to 1500 °C heat-treatment temperature, above this temperature the modulus continues to rise, whereas the strength actually decreases. There are, therefore, two main qualities of material sold: type I fibre, heat-treated to temperatures above 2500 °C, which has the higher axial stiffness but inferior strength, and type II fibre, heat-treated to only 1500 °C, which has the higher strength but rather inferior stiffness.

Other manufacturers, including Celanese and Great Lakes, start with 'dog-bone' cross-section PAN fibres and, of course, end up with carbon fibres of the same shape. It is a characteristic of such material that the tensile strength continues to increase with the modulus at temperatures above 1500 °C, which gives it some advantage over the circular cross-section fibre.

The properties are summarized in table 4. It should be noted that in all fibres there is a wide variation in strength and stiffness over the length of the fibre. Typically, type I fibre varies from 340 to 420 GN m^{-2} in axial modulus and between 1700 and 2300 MN m^{-2} in strength.

TABLE 4   *Typical properties of commercial carbon fibre derived from polyacrylonitrile*

Fibre	Manufacturer	Diameter (μm)	Density (g cm^{-3})	Average tensile modulus (GN m^{-2})	Average strength (MN m^{-2})
Round section:					
Grafil A	Courtaulds	8.0	1.74	210	2000
Type II	General	7.5	1.76	260	2500
Type I	General	7.5	1.95	380	2200
Dog-bone section:					
4T	Great Lakes		1.78	260	2400
5T	Great Lakes	5 × 13	1.85	330	2700
6T	Great Lakes		1.90	400	2800

All fibres are supplied, typically, in the form of twist-free tow containing 10000 individual filaments, each over 1 m long. Fibres can now be made continuously and so continuous-filament tow can be purchased in unspliced lengths of up to 900 m – the limit is only set by the length of continuous precursor fibre available. The actual process is secret, but it probably involves the continuous feed of yarn from one furnace to another via mercury traps, varying tension and feed speed automatically at each step. Most fibres are also treated, probably by local surface oxidation, to increase bonding with resin in carbon-fibre-reinforced resins. These composite materials will be described in chapter 8.

Further improvements in strength have been proposed by spinning the PAN precursor in dust-free conditions to produce carbon fibre without internal flaws (Moreton, 1970). However, as yet, no fibre has been made with an average tensile strength much greater than that of commercial type II fibre.

Improvement in axial modulus can be achieved by hot-stretching PAN based fibres at temperatures above 2500 °C. Johnson (1970) has shown that a Young's modulus of 700 $GN\,m^{-2}$ can be obtained by stretching PAN based fibres at temperatures between 2500 and 2800 °C. The required strain was only 30 %, which is considerably lower than the 400 % required for unoriented cellulose based fibres.

The mechanical properties will be discussed in greater detail at a later stage (chapter 6).

4 Structure of polymeric carbons

4.0 Generalities

We have already stated that when polyvinyl chloride is carbonized, it moves into a plastic state in which aromatic lamellae can rearrange themselves to form spherulites. These spherulites are really liquid crystals, or mesophase, which on further growth, will coalesce to form large continuous anisotropic grains which are oriented in response to stresses involved in the growth and movement of gas bubbles. This morphology is retained even when the coke is graphitized, the only change being the removal of boundaries marking the position of the misfit regions between aromatic layers in the original liquid crystal.

A polished surface of a graphitized or graphitizable material is revealed in an optical microscope to be highly textured. The texture and the anisotropy in various grains are clearly brought out by using polarized light (fig. 12(a)). Polycrystalline graphitized material can be seen in this way to consist of grains of various dimensions depending on the population of grain growth inhibitors, e.g. fine carbon black particles.

(a)

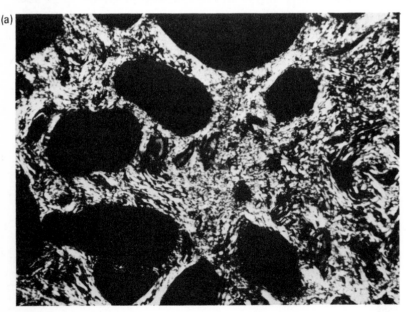

(b)

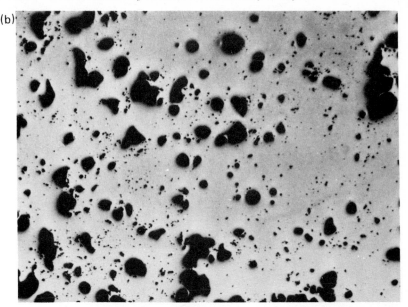

Fig. 12. (*a*) Optical micrograph of a crystalline coke derived from polyvinyl chloride. (*b*) Optical micrograph of a polymeric carbon derived from polyvinylidene chloride. Magnification ×100.

Continuity is apparent throughout each grain, though there is clearly much distortion as flow lines or striations in each coke particle ripple in an almost organic way along the cross-section. It is thought that these striations mark out lines of microcracks caused by anisotropic cooling of the individual grains in an isotropic aggregate.

When isotropic glassy carbons are polished (cf. fig. 12(*b*)) there is no visible structure or texture. Pores are apparent in the simply heated material, but these are obviously the remains of gas bubbles formed prior to carbonization. Polished cross-sections and scanning electron micrographs of charcoal show it to be quite isotropic, although retaining much of the detailed morphology of the original wood (fig. 13). Carbon fibres also retain the morphology imposed on the original precursor during the spinning stage (fig. 14).

4.1   The structure of carbons as revealed by X-rays

In order to determine the arrangement of carbon atoms in space, it is essential to study the diffraction of X-rays. A crystalline material yields a clear diffraction pattern. Single crystals of graphite, for instance, yield a hexagonal pattern of diffraction spots representing the (*hk*0) lattice

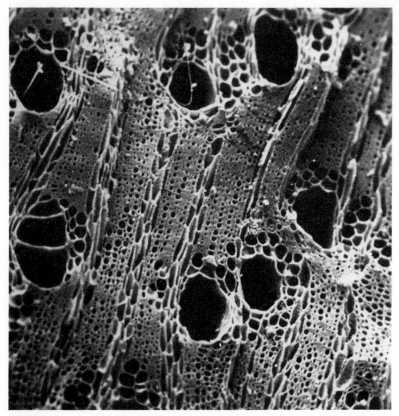

Fig. 13. Scanning electron micrograph of a charcoal, demonstrating the retention of wood morphology. Magnification ×100.

spacings when X-rays impinge along the c-axis. When X-rays pass through a polycrystalline mass of perfect graphite crystals, cones of diffracted rays are produced at various angles for all possible (hkl) lattice spacings. The fineness of the cones and the preciseness of the angle depend on the extent of periodicity and the crystal perfection.

A diffraction pattern calculated for a mass of perfect crystals, plotting intensity against diffraction angle, is displayed in fig. 15. The relevant lattice spacings are indexed; the major spacings at the lowest angles are illustrated in fig. 15. The peak of highest intensity is derived from the spacing between extensive graphite sheets and is designated (002) with a Bragg spacing of 3.355 Å. This is discernible in all carbons. The second peak (100) is related to the spacing between carbon atoms within a sheet and is observable in all annealed carbons. The strong peak at (101) is related to the extent to which extensive graphite sheets are in *ABA*

Fig. 14. Scanning electron micrograph of carbon fibres. Magnification ×2000.

register. Very little deformation is required to remove this peak; in glassy carbons (fig. 16(*b*)) it is quite absent even after high temperature annealing. Only carbons which show a definite (101) peak can really be defined as graphites.

The diffraction pattern of polymeric carbons is not sharply drawn since lattice spacings are not so well defined. The (002) peak is broader and displaced to smaller diffraction angles. The (100) peak is broader but stays very close to the perfect value. The (101) peak never appears.

Low temperature cokes and chars contain many grown-in point defects which will introduce much disorder into the pattern. Annealing at graphitizing temperatures will remove such defects but the overall morphology of the graphite sheets will remain because of the difficulty of carbon transport even at 3000 °C. Such limitations on the extent and

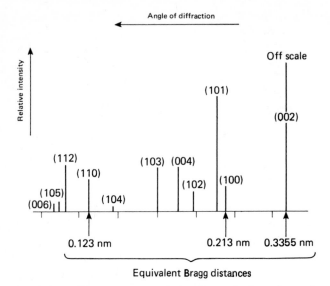

Fig. 15. Diffraction pattern calculated for perfectly random polycrystalline hexagonal graphite (after Gay and Gasparoux, 1965).

planarity of graphite sheets will leave a residual broadening which it is impossible to remove with simple heat-treatment.

It would be wise first to discuss the physical significance of the parameters which are calculated by analysing X-ray diffraction patterns and are widely used by researchers in carbon science without careful consideration. One such parameter is commonly termed 'crystallite size', a crystallite being the unit of structure in which atoms are arranged in a characteristic, periodic array with respect to orthogonal axes. Crystallite size is the apparent average size of crystallites as measured by X-ray diffraction methods, taking a cylindrical stack of graphite discs as its basic model.

This concept is based on the pioneering work of Warren (1941) who developed a method of calculation of crystallite size and applied it to structural changes in carbon black. The crystallite sizes are calculated using the Scherrer equation, $L = K\lambda/\beta\cos\theta$, where K is a shape factor, λ is the wavelength of the X-ray, β is the intrinsic breadth and θ is the Bragg angle. The mean layer diameter measured in the a-direction, L_a, is usually determined from the (110) line, whereas the average thickness measured in the c-direction, L_c, is most commonly measured from the (002) line. The shape factor, K, can have a range of values. In calculations of L_c from the (002) line, a value of 0.9 is commonly used. For random layer structures when only two-dimensional lattice reflections of the type ($hk0$) are observed, Warren (1941) has shown that a value of $K = 1.84$ should be used to calculate the layer diameter, L_a. Polycrystalline graphites lie between an oriented three-dimensional and a completely random two-dimensional structure and thus the value used to calculate L_a will have a value between 0.9 and 1.84.

Since Warren first performed his calculations of crystallite size, much work has been carried out on the structural changes in carbonaceous materials using such calculations. For instance, Franklin (1951) carried out X-ray diffraction studies of graphitizing and non-graphitizing carbons, for which she proposed various structural models. It was believed that the calculated crystallite sizes could give the absolute values of the average thickness and layer diameter of crystallites. However, it was found that the layer diameter in pyrolytic carbons as calculated using Warren's method was one order lower than that determined with an electron microscope. Thus it became obvious that the calculated crystalline size, L_a, did not coincide with the (true) value of the layer diameter. On the other hand, no significant discrepancy has been reported between the calculated crystallite size, L_c, and the average thickness observed in graphitic crystallites. Many attempts have been made to explain the discrepancy between the calculated L_a and the observed layer diameter. Diamond (1959) showed that a Gaussian strain distribution in an aromatic layer would considerably reduce the average layer diameter determined by X-ray diffraction. In general, if the imperfections are due to elastic strains, such as bending or twisting, all that may be said about the average layer diameter is that it represents a lower limit to the actual size of the imperfect sheet and an upper limit to the extent of the perfect region of the layer.

In relation to these considerations, Ergun (1969) proposed a new term, 'defect distance' instead of 'crystallite size', based on the concept of a distribution of defect concentration rather than particle size. He applied his theory to explain the structure of highly oriented carbon fibres and drew support from the experimental diffraction patterns.

The calculated crystallite diameter does not give an absolute value of the average layer diameter of graphitic stacking but it is still a useful parameter in determining the nature of carbonaceous materials and its relative value is significant. It is quite meaningless to seek to determine true crystallite sizes without knowing the nature of the strained structure in the polycrystalline material.

4.2   The process of graphitization

It is relevant to our commentary at this stage to refer to the process of graphitization because much of our knowledge of the subject relies on X-ray evidence.

Graphitization is the process by which randomly stacked, defective graphite sheets are converted into perfectly stacked (defect-free) graphite, as witnessed by the appearance of a strong (101) peak in the X-ray diffraction pattern. The process takes place simply by heating to temperatures above 2500 °C. However, in polymeric carbons, because of the short distance between boundaries, the boundary restraint is extremely great, and so perfect stacking does not occur.

Alternatively, graphitization can occur via a metal catalyst which typically dissolves the carbon atoms and allows them to reform as extensive graphite sheets which can stack perfectly in the absence of boundary restraint.

Franklin (1951) found that some non-graphitizing polymeric carbons, annealed between 2000 and 3000 °C, contain a small proportion of graphitized carbon. In graphitizing carbons, there is a single solid phase throughout the heat-treatment. When graphitic carbon is formed in non-graphitizing carbons as isolated crystals embedded in an isotropic matrix, it appears as a separate phase in the X-ray diagrams. The (002) band, in particular, is composite and either one or two fine lines are superimposed on this diffuse band. Franklin termed these effects two-phase and three-phase graphitization. She found that the small proportion of graphitic carbon formed in certain non-graphitizing carbons is more highly graphitized than in a true graphitizing carbon heat-treated under the same conditions. She concluded that the formation of a small fixed proportion of graphitic carbon in a non-graphitizing carbon was due to the high internal stresses set up during thermal treatment. However, she pointed out that the absence of strong edge-to-edge bonding is not in itself a condition sufficient for graphitization to occur at high temperatures because carbon blacks, which are also non-graphitizing, are considered to have no such bonds between neighbouring crystallites. She concluded that in order that graphitization should occur in a given carbon, the system of edge-to-edge bonding which unites the crystallites in the mass should not be too strong, and neighbouring crystallites should be suitably oriented with respect to one another.

It would be expected that similar phenomena of multi-phase graphitization would be observed in carbons heat-treated at sufficiently high temperatures under pressure. In fact, Noda and Kato (1965) showed that applied stress of the order of 1 GNm^{-2} considerably accelerates the degree of graphitization in various carbons and two-phase graphitization is observed both in graphitizing and in non-graphitizing carbons.

Mrozowski (1956) discussed graphitization mechanisms from a

different point of view. He proposed that the gradual growth of crystallites with increase of heat-treatment temperature was really ordering caused by internal stresses created by the anisotropic thermal expansion of crystallites.

Tsuzuku (1960) developed Mrozowski's qualitative explanation in a mathematical way, proposing that lateral extension of crystallites involves the movement and consolidation of tilt boundaries which are represented by vertical alignments of edge-type dislocations. He also suggested that the three-dimensional graphite structure could be due to the removal of screw dislocation grids at a twist boundary. The driving force required for the crystallite rotation resulting from this process is estimated to be comparable to the internal stresses due to the anisotropy of thermal expansion of crystallites.

The activation energies characteristic of the graphitization process are close to those for intrinsic vacancy formation and movement within the carbon sheets, and so vacancy climb of sessile defects grown into the sheets is a likely mechanism for their removal at graphitization temperatures.

We would consider that polymeric carbon is not graphitizable because, at the very onset of carbonization, a ribbon network is established. Regrouping to form extensive sheets is unlikely because this would involve the migration of carbon atoms, which requires much more energy than vacancy movement.

It is doubtful if graphitization should be promoted in a polymeric carbon fibre. It would not necessarily increase the preferred orientation of carbon–carbon bonds parallel to the fibre axis, and would lower stiffness and strength generally by introducing planes of weakness. Isolated graphitization in a polymeric carbon matrix, as observed by Franklin, would definitely introduce internal strain; the stress would be raised in the immediate vicinity of the occluded crystalline volume, leading to a drastic decrease in strength.

4.3   The changes of structure during the carbonization process

X-ray diffraction has been used to follow the changes in structure during carbonization in polymeric systems. The original precursors are typically polymeric glasses with a structure very difficult to analyse.

In all polymer glasses each atom possesses permanent neighbours at definite distances and in definite directions, although these vectorial properties relating an atom to its atomic environment are not the same for any two atoms in the assemblage. Lack of structural regularity removes from the pattern any further differentiation of the scattering in a directional sense, and this affects the available intensity information, permitting the determination of the magnitudes of the intermediate vectors but not their direction. The results can be portrayed as a radial distribution function which specifies the density of atoms (or electrons) as a function of the radial distance from any atom (or electron). The simplicity and diffuseness of non-crystalline patterns preclude the setting up of any general scheme of identification such as has been done with crystalline powder patterns. However, when the number of possibilities is limited and only pure species are encountered, the positions and intensities of the two or three discernible maxima are often uniquely characteristic of the several compounds. The positions of the peaks are calculated from the observed angles, using the Bragg equation as if reflection took place for sets of planes in a crystalline lattice. This is basically incorrect, but the equivalent crystalline d-spacings nevertheless are useful for characterizing a non-crystalline pattern.

Non-crystalline (broad) maxima actually denote the frequent occurrence of particular interatomic distances in a largely disordered substance. These distances are given in only an approximate sense by the Bragg equation. More accurate values can be obtained by analysing the phase function. It is found by simple reasoning, that the frequently occurring interaction distance R responsible for a strong maximum in the diffraction pattern is equal to 1.22 times the d-spacing calculated using the Bragg equation, i.e., the 'Bragg distance' (Klug and Alexander, 1954).

The simplified analysis ignores the effect of all other values of R in the atomic position, and height of the maxima in the scattering curve. However, it does have considerable utility in assessing the significance of the most intense maximum of a non-crystalline pattern, especially if this is far stronger than the others. In long-chain polymers, the innermost peak is nearly always the strongest and it is due principally to interatomic vectors between adjacent chains.

It is important to study the low temperature pyrolysis of the system by X-ray diffraction to throw light on the structure of the final carbon and to elucidate how the system is changed during the initial stages of carbonization prior to graphitization.

Figure 16(*a*) shows the variation of the X-ray diffraction profiles of the original phenolic resin after being carbonized at low temperatures in the carbonization regime up to 500 °C. The X-ray diffraction profiles of the original resin show a very broad and asymmetric line at a scattering angle of about 20°. This diffraction band can be approximately divided into two components, each symmetrical about its peak, the Bragg distances of which correspond to 4.6 and 3.3 Å respectively, the former

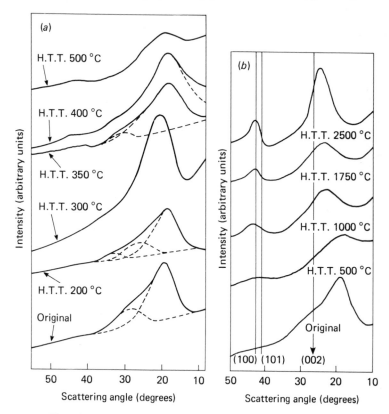

Fig. 16. X-ray diffraction profiles of (a) a phenolic resin pyrolysed progressively to 500 °C and (b) a phenolic resin char annealed progressively up to 2500 °C.

giving the higher intensity. The Bragg distance of the high-intensity peak increases to 4.7 Å after heat-treatment at 200 °C. The diffraction profile becomes sharper and more symmetric after heat-treatment at 300 °C, the Bragg distance reducing to 4.3 Å.

Infra-red spectroscopy shows that the chemical structure of the resin does not change significantly up to a heat-treatment temperature of 300 °C. Therefore, the X-ray diffraction pattern of the material heat-treated below 300 °C is the same as that of phenol–formaldehyde chain polymer.

Using the simple analysis of Klug and Alexander (1954), the approximate interchain separations are calculated to be 5.6 Å for the original resin, 5.7 Å for the resin heat-treated at 200 °C, and 5.2 Å for the resin heat-treated at 300 °C.

At a heat-treatment temperature of 330 °C, the X-ray diffraction peaks become broader, more asymmetric, and the Bragg distance of the high-intensity peak rises to 4.7 Å. It is difficult to interpret this peak. Infra-red spectroscopy suggests the formation of intermolecular cross-links at 350 °C. The chain polymer is converted into a narrow ribbon of aromatic molecules linked with carbon–carbon bonds. The strong peak may be due to the separation between two narrow ribbons, in which case the approximate separation is calculated to be 5.7 Å using the Klug–Alexander analysis. However, the peak may be due to the stacking of the narrow ribbons, in which case 4.7 Å is considered to be the interlayer spacing.

At a heat-treatment temperature of 400 °C, the diffraction profile is also broad and asymmetric but, in addition, a weak peak appears at a scattering angle lower than that of the strong peak.

At 500 °C, the diffraction profile shows much small-angle scattering and two broad peaks are resolved, the Bragg distances of which are 4.8 and 2.1 Å respectively. The peak at 2.1 Å can be detected in the resin heat-treated at 350 °C but the intensity is very weak. This diffraction profile is essentially preserved throughout the higher heat-treatment temperatures (cf. fig. 15(b)), except that a new peak is observed at a scattering angle of about 80° after pyrolysis at 1000 °C; another peak appears at a slightly higher scattering angle on annealing at 2000 °C. These peaks become sharper and shift their positions progressively to higher scattering angles with increase of heat-treatment temperature. It appears, therefore, that the structure of the final carbon is essentially only a minor modification of the structure of material produced at 500 °C.

The small crystallite size suggests that three-dimensional ordering does not take place in the system under ordinary heat-treatment conditions. In fact, no (hkl) lines other than (002) and $(hk0)$ can be detected by X-ray diffraction, even after the most prolonged annealing.

The lattice parameters and crystallite sizes calculated above are comparable to those of commercial glassy carbons reported by Fischbach and Kotlensky (1965) and Noda and Inagaki (1963, 1964).

It is implied that the material heat-treated at 500 °C consists of narrow ribbons of condensed aromatic molecules which are tangled and coiled in a complicated manner; these ribbons are linked with modified methylene bridges, resulting in a three-dimensional network structure. Pyrolysis above 500 °C involves the removal of hydrogen, and the internal strain energy in the narrow ribbons is thereby gradually released. However, this release is restricted by strong bonds between the ribbons. The broad X-ray diffraction peak with the Bragg distance of 4.8 Å in the system heat-treated at 500 °C is identified with the stacking of the ribbons which are locally overlapped and, therefore, 4.8 Å is an average inter-layer spacing between the ribbons. This value decreases continuously with increase of heat-treatment temperature.

Strong small-angle scattering detected in the X-ray diffraction profiles of the system heat-treated at above 500 °C indicates that the material contains a large proportion of fine pores. This is because tangled and highly strained ribbons cannot fill space efficiently. Optical microscopy cannot detect any macroscopic pores and there is no evidence of texturing throughout the heat-treatment process. Polarized light microscopy and X-ray diffraction texture studies do not reveal anisotropy. It is inferred that the narrow ribbons of condensed aromatic molecules are not drawn straight and parallel in a certain direction but are quite randomly oriented.

A conventional analysis of X-ray diffraction profiles can be carried out for annealing temperatures above 500 °C, assuming that the carbon has a graphitic structure. Table 5 shows the variations of average lattice parameters ('Bragg distances') and crystallite sizes calculated using the modified Warren equations developed by Short and Walker (1963). These should be compared with typical values obtained for a graphitizable coke, for which d-spacing = 3.44 Å, $L_c = 20$ Å, $L_a = 90$ Å after heat-treatment at 1000 °C, and d-spacing = 3.35 Å, $L_c = 500$ Å and $L_a = 2000$ Å after heat-treatment at 2700 °C.

The interlayer spacing calculated from (002) bands in polymeric carbon is very large; it is larger than 3.44 Å even after annealing at 2700 °C. 3.44 Å is considered arbitrarily to be the average interlayer spacing when extensive graphitic layers are randomly stacked. The

TABLE 5   *Variation of 'crystal' parameters with annealing temperature*

H.H.T. (°C)	d (Å)	a (Å)	L_c (002) (Å)	L_a (100) (Å)	L_a (110) (Å)
500	4.80	–	12	–	–
700	4.11	–	13	–	–
900	3.88	–	14	27	–
1000	3.89	2.41	14	29	–
1250	3.90	2.41	15	31	–
1500	3.88	2.41	15	35	35
1750	3.75	2.42	16	38	35
2000	3.58	2.43	22	53	39
2500	3.56	2.43	24	54	48
2700	3.49	2.43	30	60	49

interlayer spacing decreases to 3.35 Å for perfect stacking. This suggests that the carbon layers are highly strained and the internal strain energy is not released by simple heat-treatment. The lattice parameter along the *a*-axis is significantly smaller than that of a single crystal of graphite. This also indicates strain in the carbon layers. The carbon layers in the crystallite are apparently compressed along the basal plane and bent by the surrounding crystallites. The non-graphitizability of the system suggests that the configuration of the units of the ribbons is a complicated three-dimensional one which inhibits the thermal release of the strain energy in the carbon layers.

The apparent crystallite sizes, calculated from the (002), (100) and (110) bands are much smaller than those of graphitizing carbons. The apparent layer diameter, L_a, is less than 100 Å even after annealing at 2700 °C.

X-ray parameters in oriented carbon fibres derived from PAN have been calculated by Johnson and Tyson (1970) for a range of heat-treatment temperatures. Up to 2000 °C, the apparent crystallite dimensions are close to those for our glassy carbon. Above 2000 °C, however, there is a substantial increase in dimensions. At 3000 °C, both L_a and L_c rise to about 90 Å. That is, of course, still much less than the values in graphitized material. It is inferred that the alignment of microfibrils allows a limited increase in their dimensions but does not necessarily promote graphitization.

4.4   Small-angle scattering

Many carbons show pronounced scattering of X-rays at small angles attributable to fluctuations of the electron density on a scale larger than that of interatomic distances (Ruland, 1969). Earlier studies of Franklin (1951) showed that strong small-angle scattering was present in polymeric carbons. Her evaluation in terms of a radial density distribution is not really valid and the numerical results must be viewed with caution.

A more generally accepted treatment, due to Guinier and Fournet (1955), is to plot the logarithm of the X-ray scattering intensity (I) against the square of the scattering angle (ε^2). The slope is given by $-4\pi^2 R^2/3\lambda^2$, where λ is the X-ray wavelength and R is the radius of gyration of scattering particles or voids. This law is obeyed by systems containing particles or voids of approximately equal size in a homogeneous matrix.

Such measurements have been made by A. C. Craievich of the University of São Carlos, Brazil, using disc specimens of our phenolic resin

TABLE 6   *The variation of pore size and density with temperature of pyrolysis of phenolic resin*

Heat-treatment temperature	Density (gcm^{-3})	Radius (Å)
500	1.14	–
600	1.17	4.2
700	1.26	4.7
800	1.43	7.0
900	1.52	7.9
1000	1.55	8.4

pyrolysed at various heat-treatment temperatures between 500 and 1000 °C. He finds that in all cases a linear portion of the Guinier plot is measurable between $\varepsilon = 1$ and $\varepsilon = 2.3°$. For $\varepsilon < 1°$ the slope of every curve increases rapidly either because of the presence of larger voids with diameters greater than 500 Å or, perhaps, because the voids are very flat oblate ellipsoids.

Table 6 shows the variation of the radius of pores (or regions of high density) with heat-treatment temperature, together with the density. It is clear that increases in R correlate with increases in density, possibly because dehydration draws together the microfibrils during this pyrolysis regime, eliminating a substantial population of small voids to leave a residual population of voids with an average diameter of 8.4 Å.

Maxima in the small-angle scattering of carbonization residues of well-defined organic compounds have been observed and are attributed to short-range order in the side-by-side packing of layers of nearly identical dimensions (Ruland, 1965).

Small-angle scattering from high modulus carbon fibres is also pronounced and shows the same degree of preferred orientation that is indicated by (002) arcs (cf. section 4.7). It is inferred that the pores are elongated and lie parallel to the component crystallites (or microfibrils). Detailed investigations of these small-angle scattering patterns have been carried out by Perret and Ruland (1969) and Johnson (1971).

4.5   The structure of polymeric carbons

It is inferred from our carbonization studies that glassy carbon is made up of condensed aromatic ribbon molecules which are orientated randomly and tangled in a complicated manner. The high resistance to graphitization of glassy carbon suggests that the configuration of these ribbons is very stable. The main questions on the structure of glassy carbon are whether the structure is two-dimensional as in graphite or three-dimensional as in glass and whether the fully formed glassy carbon contains tetrahedral carbon atoms in addition to trigonal carbon atoms.

The possibility that atoms are present in diamond-like structures of tetrahedral carbon atoms has been discussed by Ergun and Tiensuu (1959), Ergun and Alexander (1962) and Kakinoki *et al.* (1960). The presence of such structures is taken as an explanation for the extreme resistance of glassy carbons to graphitization. The main argument put forward by Ergun and his co-workers is that since interference functions of very small diamond-like domains are similar to those of small graphitic layers, their presence in carbons with small apparent crystallite sizes cannot be excluded. The coincidence of the interference functions is limited to the first ($hk0$) lines of the random-layer structure. No interference is produced near the (002) line which is always observed together with the ($hk0$) lines even in highly disordered carbons. A detailed study of the relative intensities of the (002) and ($hk0$) lines, taking into account intensity loss due to disorder defects, has not been made to check whether diamond-like structure is really present. Noda and Inagaki (1964) calculated the radial distribution function for glassy carbon and assumed the presence of both trigonal and tetrahedral atoms. This was similar to Kakinoki's work based on electron diffraction studies at very large Bragg angles where only shorter interatomic distances give appreciable contributions to the interference function. Assuming only next-neighbour distances are effective, Kakinoki found that the observed intensity distribution could only be fitted to that calculated if two nearest-neighbour distances at 1.55 and 1.41 Å are assumed with equal probability – i.e., that diamond and graphite are present in equal amounts. The assumption that only next-neighbour distances are important at larger Bragg angles is not valid; it is expected that the second and third neighbour still has a well-defined distance even in highly disordered carbon.

Furukawa (1964) criticized the above in view of the large difference between the observed density and that calculated from a high proportion of tetrahedral carbon. He proposed a 'cubic' model with an average co-ordination number for each atom and a mixture of single, double and triple bonds to produce a three-dimensional network.

Kakinoki (1965) modified the picture of Noda and Inagaki with the addition of oxygen bridges between trigonal and tetrahedral carbon volumes. However, the large amount of oxygen in glassy carbon was assumed from data on moist carbon and this needs to be substantially corrected. It is difficult to believe that extensive oxygen bonding could persist above 2000 °C.

Takahashi and Westrum (1970) measured the specific heat of glassy

carbon heat-treated at 3000 °C and they observed that the specific heat
is proportional to the square of the absolute ambient temperature at low
temperature, indicating that the structure of glassy carbon is essentially
two-dimensional.

Halpin and Jenkins (1969) studied alkali metal attached on glassy
carbon and found that intercalation compounds are formed in the
carbon after exposure to potassium vapour. This confirms that glassy
carbon has a layer structure and demonstrates that the X-ray diffraction
line observed at a scattering angle of about 26° is due to the stacking
of aromatic molecules in vitreous material.

If the magnification and resolving power of the electron microscope
are sufficiently high, the essential layer structure in glassy carbon can
be observed directly. Ordinary low resolution transmission electron
micrographs of a typical glassy carbon show that the edges consist of a
network arrangement of strings or microfibrils, the thickness of which is
about 30 Å. Each string is considered to be a stack of graphite-like
ribbon molecules. The thickness of each string is similar to L_c, determined by X-ray diffraction. It is also possible to discern micropores,
created when the strings form a network structure. The diameter of
the pores is estimated to be between 50 and 100 Å, comparable to that
calculated from small-angle X-ray scattering.

Using a phase-contrast technique, high resolution electron microscopy
studies on carbons derived from phenolic resin were carried out (Jenkins
et al., 1972). Figure 17(*a*) shows a high resolution electron micrograph of
a pyropolymer heat-treated at 500 °C. In the main, only random short-
range order exists. The structure of the same material heat-treated at
900 °C is revealed in fig. 17(*b*). This shows that there is short-range
order with two or three parallel wrinkled layer groups which are randomly orientated.

Figures 17(*c*) and 18 are micrographs of the material annealed at
2700 °C. The system is made up of stacks of two-dimensional ribbon-like
molecules, reminiscent of fibrils in crystalline polymers. The fibrils are
randomly orientated and tangled in a complicated manner. The height
of a stack of the ribbon-like molecules is estimated to be up to 40 Å.
The value agrees well with L_c calculated from X-ray line broadening.
It is difficult to estimate the length of the ribbon molecules from this
micrograph because they do not show clear-cut boundaries and continue
through many crystallites. It is possible, however, to estimate the
distance between high-angle misfits, giving the length of relatively
straight parts of the molecules. This is reckoned to be 100 Å, which is
comparable to L_a calculated previously from X-ray diffraction studies.
Branching of the graphite-like molecules is clearly observed.

The structure of fig. 17(*c*) is characteristic of the surface, showing

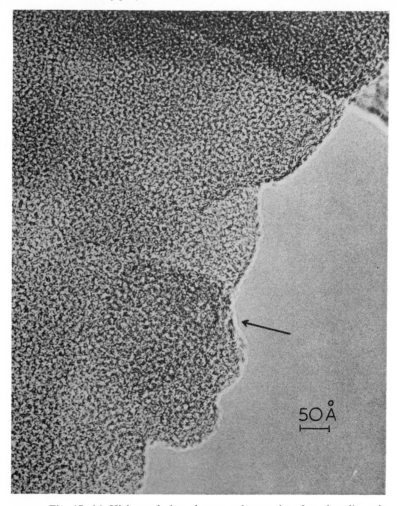

Fig. 17. (a) High resolution electron micrographs of a phenolic resin carbonized at 500 °C.

continuity of microfibrils to form loops projecting from the surface. Orientation is clearly random and fracture occurs at the interfibrillar boundary. *No 'loose ends' are apparent*; *this explains the characteristic chemical inertness of glassy carbon.* In the interior of the fragment the material retains its essential randomness and microfibril continuity. There are indications that the width of the layers, both perpendicular and at different angles to the layer images, can be large locally. This must be attributed to the presence of many confluences in the ribbon

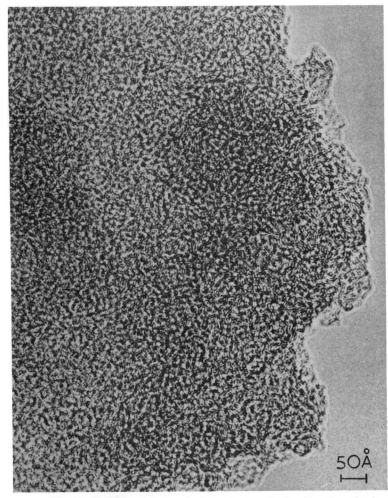

Fig. 17. (*b*) High resolution electron micrograph of a phenolic resin char
heat-treated at 900 °C.

network where neighbouring ribbons merge to form a much wider ribbon
locally.

Figure 18 shows another micrograph of polymeric carbon derived
from phenolic resin fibres, subsequently annealed at 2700 °C. Here,
preferred orientation of the microfibrils is clearly observed. It is also
obvious that there is a continuous transition from random to
orientated fibrils. There is no abrupt phase change, and the two forms
consist essentially of the same material.

A three-dimensional structural model for isotropic glassy carbon is

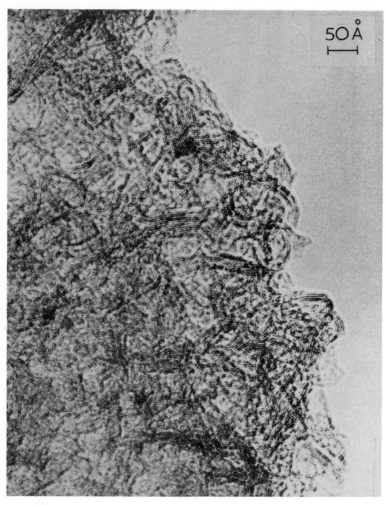

Fig. 17. (*c*) High resolution electron micrograph of a phenolic resin char heat-treated at 2700 °C.

illustrated in fig. 19. This is highly simplified and, in reality, the fibrils should be packed closer together to form a tighter mass. We have also drawn in our conjecture of 'confluences' in the network of microfibrils. The strong confluences occur where ribbons merge into each other; weak confluences occur where ribbons overlap.

Fig. 18. High resolution electron micrograph of a phenolic resin char heat-treated at 2700 °C, showing preferred orientation.

We may take it from the mass of information presented in this book, that polymeric carbons are produced by the progressive coalescence of stabilized polymer chains in the temperature region between 400 and 1000 °C. This defines the carbonization process. The final structure is laid down at the start of the coalescence, but is only attained progressively as the temperature is raised because of the increasing difficulty in ordering as the coalescence proceeds. This difficulty can be attributed

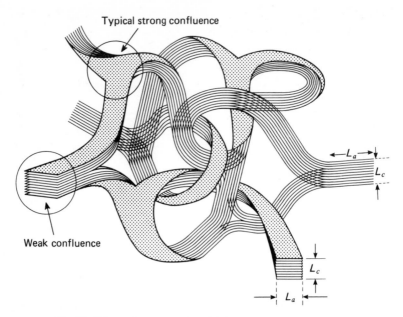

Fig. 19. Schematic structural model for a glassy carbon.

to the competition between adjacent polymer chains, thus hampering the attainment of the final graphite ribbons.

At 1000 °C these ribbons possess high local distortions due to internal defects and bonds with adjacent ribbons which probably exist at these temperatures both perpendicular and parallel to the plane of each ribbon. As the carbons are heated above 1000 °C these local defects are gradually eliminated until at 2700 °C only perfectly smooth ribbons remain. These stack above each other with up to 10 ribbons in each stack, forming a microfibril. The microfibrils are bonded to each other at the edges of the ribbons with strong covalent bonds which prevent the narrow graphite ribbons from moving into *A–B–A* stacking register and so no three-dimensional graphite lines are possible. The microfibrils twist, bend and intertwine. In glassy carbon, they are randomly arranged in space. The geometry is such that pores are inevitable between the microfibrils. After heat-treatment at 2700 °C all the carbon atoms are in the sp_2 state; there is no unambiguous evidence for carbon atoms in the sp_3 state.

The development of parallel stacked extensive sheets of graphite is prevented because continuity is preserved along the length of each ribbon by strong C–C bonds, which would have to be broken to produce extensive areas of graphite sheet. The distortions allowable in narrow

ribbons with edge-to-edge bonds are much greater and much more
varied than those allowed in sheets; twisting, bending, warping and
weaving are all easily maintained in ribbons, but not in sheets. The
large area free from constraining bonds in graphitic materials gives them
a higher strain to fracture compared with the more brittle isotropic
polymeric materials.

Incidentally, it should be noted that there has always been the opinion
that all polycrystalline carbons can be regarded as polymer systems in
which sheets of graphite are bonded to sheets in neighbouring crystals
with fractional double bonds and tetrahedral-type linkages. Thus, as
Mrozowski (1956) first said, 'a carbon solid is essentially a three-dimensional network of crystallites held rigidly by a complicated system of
cross-links which stabilize the structure to such an extent that parallel
shifts of planes in crystallites can occur only under forces sufficient to
break (or distort) the peripheral C–C bonds'.

4.6   The structure of carbon fibres

Yamaguchi (1964) measured the electrical properties of carbonized
polyacrilonitrile fibres. The behaviour of electrical resistivity, magneto-
resistance and thermo-electrical power of carbonized polyacrilonitrile
fibres as a function of heat-treatment temperature is similar to that of
glassy carbon and completely different from that of graphitizing carbons.
This shows that carbon fibres and glassy carbons belong to the same
group of carbon materials.

Johnson and Watt (1967) studied the X-ray diffraction of high
modulus carbon fibres derived from polyacrilonitrile and reported that
L_c comprises at least twelve layer planes and L_a lies between 60 and
120 Å. Electron microscopy indicated that carbon fibres have a network
structure of fibril units, the length of which is at least 1000 Å. Johnson
and Watt suggested that the modulus of the carbon fibre is determined
by the orientation of the graphite crystallites within the carbon fibrils,
while the strength is a function of the interfibrillar bonding.

Ruland (1969a) made an intensive study of the structure of the two
highly anisotropic carbon fibres based on PAN and rayon and considers
the two to be very similar. The absence of (*hkl*) lines even after high
annealing temperatures indicates that the fibres are not graphitizable.
The carbon atoms form two-dimensional hexagonal layers with apparent
linear extension of 60–120 Å, and these layers form small stacks between
30–100 Å thick. The general state of the layer ordering is the same as
that of bulk glassy carbons.

The only difference between bulk and PAN fibres is that carbon microfibrils are preferentially oriented parallel to the fibre axis so that the parallel stacking of the layers is preferentially perpendicular to this axis, and the basal-plane interference appears predominantly on the equator of the X-ray fibre diagram. The angular distribution of the interference varies considerably with heat-treatment temperatures (Ruland, 1967) and stretching (Johnson *et al.*, 1969). The orientation of the carbon layers improves with increasing heat-treatment temperature and stretching, the best fibres representing an average deviation from perfect orientation of only $\pm 4°$. The orientation of the two-dimensional hexagonal structure of the layers with respect to the layer normals is random.

Small-angle scattering of X-rays shows the existence of needle-like pores with cross-sections ~ 15 Å and lengths greater than 300 Å (Ruland, 1969*b*). The pore axes are oriented preferentially parallel to the fibre axis and the angular distribution of the axis is similar to that of the layer planes in that the distribution becomes narrower with heat-treatment and stretching. The sharpness of the density transition between pore and carbon suggests that the pore walls are formed by carbon planes. The pores are completely inaccessible to helium.

It has been shown that potassium and caesium intercalate easily in stiff fibres (Hérinckx *et al.*, 1968) with no change in preferred orientation of the carbon layers or the pore axes. This indicates that no cross-links occur perpendicular to the carbon layers.

Ruland concludes (1969) that oriented carbon fibres comprise long winding ribbons stacked above each other and preferentially oriented parallel to the fibre axis. The stacks form structural units, which may be termed microfibrils, which are wrinkled and twisted and must thus leave voids which are determined by the shape and orientation of the microfibrils. The average length of the carbon layers in the direction of the fibre axis deduced from X-ray profile studies is thus a measure only of the extension of planar regions of the ribbons and represents the distance between wrinkles in the ribbons. A schematic representation of this model is shown in fig. 20(*a*).

Studies of high resolution electron microscopy (Fourdeux *et al.*, 1969) using the phase-contrast technique confirm the essential features of this model and show notably that the carbon layers form continuous wrinkled ribbons greater than 2000 Å in length (cf. fig. 19(*b*)).

Polished cross-sections of PAN carbon fibres embedded in epoxy resins have been viewed by Butler and Diefendorf (1969) using incident polarized light. It was possible to discern well-defined extinction contours arranged in the form of a Maltese cross. This remains parallel to the polarizers as the specimen is rotated, indicating that a tree-trunk-type

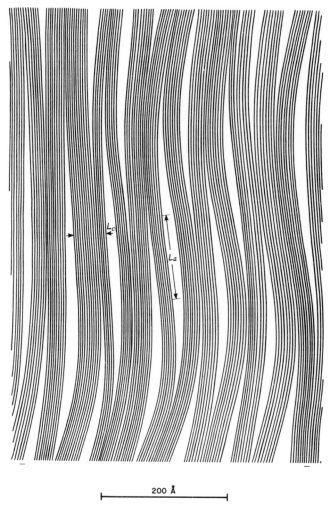

200 Å

Fig. 20. (*a*) Schematic representation of the structure of high modulus carbon fibres (Ruland, 1969).

of texture is present, with graphite ribbons connected to each other to form a system of concentric cylinders, and each stack of ribbons lying nearly parallel to the fibre surface; on larger fibres where the centres were not stabilized by oxidation prior to carbonization, there was a more pronounced structure.

Other carbon fibre sections are clearly derived directly from the morphology of the original precursor fibres, imprinted during the

Fig. 20

spinning process. 'Dog-bone' sections and fluting are all characteristics of different precursors.

Figure 21 shows a polarized light micrograph of cross-sections of carbon fibres derived from phenolic resin and heat-treated at 2500 °C. It is observed that the cross-sections of the fibres have four bright areas and four dark spots in the periphery. The bright arcs and dark arcs are spaced apart by ninety degrees. This suggests an isotropic core is surrounded by an anisotropic textured sheath with a high degree of preferred orientation. Electron diffraction studies confirm preferred orientation in the surface.

Figure 22 provides further evidence of a difference in texture between core and surface in a carbon fibre derived from phenolic resin. This scanning electron micrograph shows an end of a fibre which was fractured before carbonization, and subsequently annealed at 900 °C. The core has shrunk much more along the fibre axis than the surface. Surface

(c)

Fig. 20. (*b*) A high resolution electron micrograph showing the orientated microfibrillar structure of pitch based fibre heat-treated to 2800 °C with an extension of 140%. The fibre was manufactured by H. M. Hawthorne of the University of British Columbia and is reproduced by permission of D. J. Johnson and D. Crawford of the University of Leeds. (*c*) A high resolution electron micrograph showing the orientated microfibrillar structure of type I PAN based fibre. Reproduced by permission of D. J. Johnson and D. Crawford of the University of Leeds.

orientation in carbonized fibres may be thought perhaps to be caused by two-phase graphitization observed in many non-graphitizing carbons. However, no lines apart from (002) and (*hk*0) can be detected by X-ray diffraction even in fibres annealed at 2700 °C. This indicates that the oriented layer is not graphite but 'textured glassy carbon'. A comparison of the structure of the sheath material with that of high modulus fibres from polyacrilonitrile, which also show a high degree of preferred orientation in the absence of (*hk*0) lines in the X-ray diffraction profile, suggests that the two are identical.

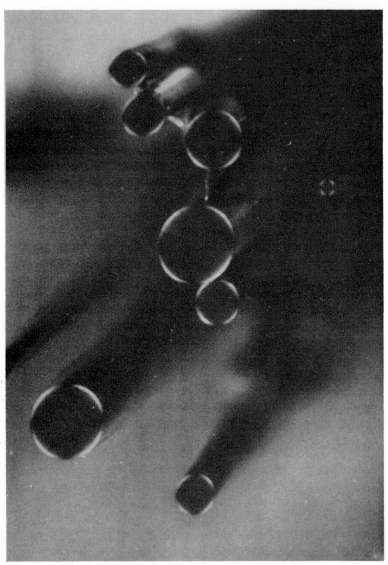

Fig. 21. Sections of 'two phase' carbon fibres derived from phenolic resins viewed under polarized light, showing outer sheaths with preferred orientation. Magnification ×800.

Since a fibre with a small diameter has a considerably larger ratio of free surface area to volume than a bulk sample, the release of strain energy in the surface should be easier than in a bulk sample. This could explain why surface orientation is observed in the fibres. If so, particles

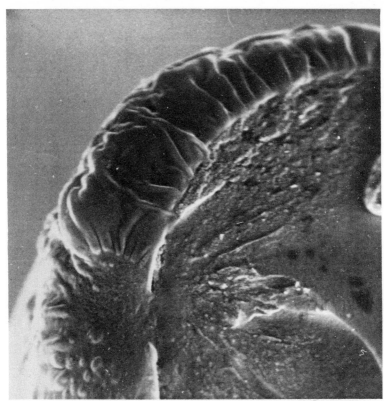

Fig. 22. Scanning electron micrograph of a phenolic resin carbon fibre fractured prior to carbonization, showing outer skin. ×3000.

made from the same starting material with a thickness comparable to the diameter of the fibres should show even greater surface orientation because the ratio of free surface area to volume is larger in particles than in fibres. However, no evidence of surface orientation is observed in carbonized finely divided phenol–hexamine particles. In addition, if this geometrical factor results in surface orientation, fibres with a smaller diameter should have larger thickness of oriented phase. The thickness of the oriented layer seems, however, to be independent of the fibre diameter. It is concluded that a free surface-to-volume factor is not responsible for surface orientation.

Our explanation for the presence of textured sheaths in fibres derived from phenolic resin is that preferred orientation is introduced during melt extrusion and subsequent spinning. The preferred orientation in the polymer chains is rendered thermally stable by oxidation in the surface. Changes in the polymer structure are observed which confirm this; in particular, an additional infra-red absorption band appears at 1660 cm^{-1}, which is a characteristic of oxidized phenolic resin. An increase in thickness of the oriented layer after heat-treatment at high temperatures can be caused by the difference in thermal expansion between core and sheath; tensile stresses are thereby imposed on the core which will tend to increase the proportion of microfibrils oriented along the fibre axis.

4.7 Preferred orientation

Even though high modulus carbon fibres are not truly graphitic, the (002) X-ray reflections are sufficiently sharp for the use of X-rays in determining the preferred orientation of component microfibrils.

The characterization of the preferred orientation of graphitic ribbons in polymeric carbons is of importance in defining anisotropy in carbon fibres.

X-ray transmission photographs can be taken through bundles of aligned fibres (Brydges *et al.*, 1969). Figure 23 shows a typical X-ray photograph taken with a bundle of PAN fibres. The (002) reflections on

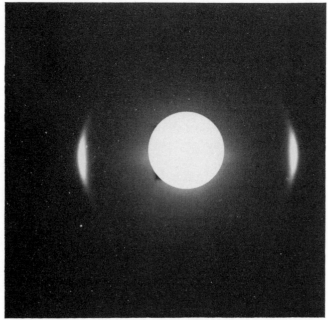

Fig. 23. X-ray photograph showing preferred orientation in carbon fibres derived from polyacrylonitrile.

the fibres can be analysed on a microdensitometer to establish the X-ray intensity as a function of the scanning angle on the fibre. The fibre angle is related to the angle (ϕ) between the bundle axis and the c-axes of the crystallites giving reflection. Ruland (1967) measured the X-ray scattering of a fibre on a frame with a counter device and a proportional counter with pulse-height discrimination. The intensity distribution of the (002) band was determined as a function of the Bragg angle for $\phi = 0$ and $\phi = \pi/2$. The maximum intensity of the line, the integral width of the line and the background under the line were determined. The intensity distribution in ϕ was then measured for a fixed Bragg angle and the background subtracted. The Fourier coefficients of the intensity curves were computed and corrected for errors due to the slit system and the tilt angle of the fibres on the frame.

Reynolds (1970) has managed to determine the orientation function for crystallites in a single fibre. An intense radiation up to 50 kV is produced from a spot 10 μm in diameter. This is brought to a focus using a gold-plated toroidal mirror so that exposures of only a few hours are required to produce good (002) diffraction arcs from single fibres.

The foregoing X-ray scattering techniques allow us to determine the angular distribution of layer normals $g(\phi)$. Typically, $g(\phi) = \sin^n \phi$. The total number of c-axis poles ($n(\phi)$) impinging on a surrounding sphere of radius R at angles between Φ and $\phi + \mathrm{d}\phi$ would be given by

$$n(\phi) = g(\phi) \cdot 2R \sin \phi \, \mathrm{d}\phi$$

The total number of c-axis poles impinging on the whole sphere would then be given by

$$\int\limits_0^{\pi/2} n(\phi) \, \mathrm{d}\phi = \int\limits_0^{\pi/2} 2 \cdot g(\phi) \cdot R \sin \phi \, \mathrm{d}\phi$$

Various derived preferred orientation parameters have been calculated from the distribution of layer normals. Preferred orientation can be defined by a parameter termed R_z by Ruland (1967) given by

$$\frac{\int\limits_0^{\pi/2} \sin^2 \phi \, n(\phi) \, \mathrm{d}\phi}{\int\limits_0^{\pi/2} n(\phi) \, \mathrm{d}\phi}$$

which equals

$$\frac{\int\limits_{0}^{\pi/2} g(\phi) \sin^3 \phi \, d\phi}{\int\limits_{0}^{\pi/2} g(\phi) \sin \phi \, d\phi}$$

Theoretically, this should serve to define the thermal expansion coefficients, as will be discussed later.

Bacon (1956) used the same parameter to define preferred orientation in cylindrically symmetrical extruded electrographite. As the preferred orientation changes from the isotropic state to the fully oriented, R_z increases from $\frac{2}{3}$ to 1.

Ruland (1967) was able to examine the change in R_z with heat-treatment temperature. It was shown that the preferred orientation in the two starting materials, a rayon yarn and an acetate cellulose fibre, with $R_z = 0.85$ and 0.96 respectively, dropped suddenly, on heating between 240 and 280 °C, to the value for random orientation ($\frac{2}{3}$). A small but significant preferred, orientation reappears after heat-treatment above 1000 °C and increases with higher heat-treatment up to 2800 °C to reach R_z values of 0.85 in the case of the rayon and 0.77 in the case of the cellulose acetate. Thus, highly oriented cellulose fibres do not necessarily produce well-oriented carbon fibres. It is significant, though, that there is a distinct probability that at least some of the original preferred orientation will be remembered after carbonization.

The major interest in preferred orientation, however, is in the rapid estimation of structure-sensitive physical properties of fibres, especially elastic constants and thermal expansion coefficients, especially the values perpendicular to the fibre axis, which are difficult to measure directly. These will be dealt with in the appropriate sections to follow. It is interesting to note here that Reynolds (1970) has shown that the preferred orientations of fibres change when they are pulled; these changes are completely eliminated by removing the load.

5 Physical properties

5.0 Generalities

The physical properties of polymeric carbons are directly related to the structure. Thermal vibrations are transferred along the ribbons and the greatest amplitudes are predominantly out-of-plane. The thermal properties must, therefore, be very similar to those of normal polycrystalline graphites with small crystallite sizes. The high thermal expansion of free graphite sheets along the c-axis is restricted by the frequent edge-to-edge bonding at ribbon boundaries. Electron movement along perfect ribbons is easy even though the path is circuitous. The resistivity should, therefore, be very close to that of normal polycrystalline graphite.

The confirmation of these predictions further justifies our adoption of a ribbon model for polymeric carbons.

5.1 Thermodynamic quantities and thermal conductivity

The heat capacity of glassy carbon has been measured by Takahashi and Westrum (1970) from 5 to 350 K – it shows no transitions or thermal anomalies. In common with materials consisting of large sheets of carbon atoms, e.g. pyrolytic graphite, it follows an approximate T^2 dependence up to 30 K. This does indicate the presence of graphitic lamellae with no contribution from sp$_3$-type carbon. The actual heat capacity is slightly higher than that for a true graphite and so a certain proportion of the bonds must be weaker. Thermodynamic quantities measured at 300 K were the heat capacity (2.05 cal K^{-1} mole^{-1}) and entropy ($S - S_0 = 1.406$ cal K^{-1} mole^{-1}).

Bale (1970) measured the enthalpy of PVDC char annealed at 2500 °C between 550 and 1300 K. He compared these results with data on Ceylon graphite and PVC coke annealed at the same temperature. The enthalpies of all three do not differ much and thus each must consist of carbon atoms in exactly the same configurations.

Davidson and Losty (1963) developed a technique for the determination of thermal conductivity in thin discs of cellulose glassy carbon. Thermal diffusivity (= thermal conductivity (specific heat × density)) at room temperature was measured by heating the front surface of a disc with a standard electronic flash tube, the duration of the flash being about 1 millisecond. The heat which is initially stored in a thin surface film flows through the material at a rate determined by the thermal conductivity and the effective thermal inertia. The e.m.f. from a fine thermocouple at the back surface was recorded on an oscilloscope. Analysis of the temperature–time curve yielded the thermal diffusivity and the specific heat from which the thermal conductivity is calculated.

High temperature measurements at 1500 °C were made by Davidson and Losty (1963) using a modulated electron beam. The thermal diffusivity at room temperature is $0.04 \text{ cm}^2\text{s}^{-1}$ rising very slightly to $0.05 \text{ cm}^2\text{s}^{-1}$ at 1500 °C. Using published data for the specific heat, the thermal conductivity was calculated to be $0.01 \text{ cal cm}^{-1}\text{s}^{-1}\text{K}^{-1}$ at 25 °C and 0.04 units at 1500 °C. This fourfold increase, mainly accounted for by an increase in specific heat, is in marked contrast with the behaviour of polycrystalline graphites which have a much higher conductivity of 0.4 units at room temperature but this decreases to 0.06 units at 1500 °C. The tendency for all carbons of varying crystallinity to approach a common conductivity at elevated temperatures may be because the predominant phonon scattering process in polycrystalline graphite is Umklapp scattering. As the test temperature rises the mean free path due to this type of scattering decreases exponentially. In glassy carbon, the predominant scattering process may be that due to the boundaries. Thus as the test temperature is increased the thermal conductivity of carbons containing large crystallites approaches that of carbon containing small crystallites. Heat-treatment causes the thermal conductivity to increase from 0.01 at 1000 °C to 0.02 at 2000 °C and 0.036 at 3000 °C.

Strauss (1963) has measured the thermal conductivity of hard filler–hard binder carbons at high temperatures and has shown that such materials also show an increase in conductivity so as to approach the decreasing values exhibited by soft filler–soft binder graphites as the temperatures increase. For instance, the thermal conductivity (k) of a hard filler–hard binder material of density 1.61 gcm^{-3} has a value of $0.45 \text{ cal cm}^{-1}\text{s}^{-1}\text{K}^{-1}$. Taking into account the effect of porosity, these values are very close and should be compared with those of single crystal graphite which has a k value of $0.6 \text{ cal cm}^{-1}\text{s}^{-1}\text{K}^{-1}$ at 300 K in the basal plane and $0.17 \text{ cal cm}^{-1}\text{s}^{-1}\text{K}^{-1}$ at 300 K perpendicular to the basal plane. Polycrystalline graphites have values between 0.25 and 0.40 $\text{cal cm}^{-1}\text{s}^{-1}\text{K}^{-1}$.

5.2   Thermal expansion

The thermal expansion coefficient of isotropic polymeric carbons depends on the mode of manufacture and the starting materials, varying from 2.0 to 3.4×10^{-6} K^{-1} for linear expansion. For a given material it does not depend on heat-treatment temperature between 1300 °C and 3000 °C, but it does increase with the ambient test temperature from 2.0×10^{-6} K^{-1} to 4.0×10^{-6} K^{-1} at 2000 °C (Yamada, 1968).

These values should be compared with those of single crystal graphite for which α along the c-axis (α_c) is 27.3×10^{-6} K^{-1} and α along an a-axis (α_a) is zero at 4000 °C. Below, α_a is negative; above, it rises to saturation at 1.2×10^{-6} K^{-1}. Energy is accumulated preferentially in c-axis vibration at low temperatures which causes considerable expansion in that direction, leading to a contraction in the basal planes. At high temperatures, in-plane nodes accumulate the thermal energy causing a renewal of these dimensional changes. Polycrystalline graphite has only a fraction of the thermal expansion of single crystal graphite – between 2 and 4×10^{-6} K^{-1} at room temperature, increasing to between 4 and 7×10^{-6} K^{-1} at 2000 °C. These low figures are usually attributed to the presence of so-called 'Mrozowski cracks' (1956) due to the anisotropic contractions of individual crystals in the general matrix during cooling from the firing temperature ($\sim$2700 °C).

It is difficult to conceive of such an explanation for polymeric carbons, since the presence of such cracks would establish points of rupture in such a brittle material. The distance apart of the layers is much greater than that for single crystals of graphite. However, the graphite layers must be restrained in the opposite sense from achieving their true spacing of 3.35 Å by the same intercrystalline bonds, and it is predicted that differential internal strain will serve only to relieve already existing internal stresses. The lower thermal expansion is a clear manifestation of the presence of strong boundary restraint and weak bonds within the crystal. The thermal expansion coefficient of glassy carbons annealed at 3400 °C is still only $\frac{1}{3}$ of that of isotropic Acheson-type graphites. This does indicate that it is the presence of strong intercrystalline bonds which prevents the full strain-free c-axis thermal expansion of the single crystal from asserting its full influence on the overall coefficient.

The thermal expansion coefficient of carbon fibres is very important for calculations of thermal stresses on fibres in various matrices. The axial coefficient of thermal expansion of oriented PAN carbon fibres has been found to be -1.0×10^{-6} K^{-1} at room temperature.

Butler (1973) has measured the axial coefficient of expansion of carbon fibres between 800 and 1600 °C. In this region the a-axis expansion of a graphite sheet is 0.4×10^{-6} K^{-1}. The fibre coefficient varied from 3.9×10^{-6} K^{-1} for isotropic fibre to 0.5×10^{-6} K^{-1} for high modulus Thornel 75. Butler shows that the relationship of thermal expansion to preferred orientation determined by X-ray analysis is best explained by a fibril model, each fibril consisting of continuous, sinusoidally undulating ribbons.

The coefficients perpendicular to the fibre axis have never been determined and are the subject of some controversy. If free crystallites are assumed, as in the Bacon (1956) model, we would predict a very high coefficient of $\sim 13 \times 10^{-6}$ K^{-1}. In actual fact, the crystallites may be constrained into a tree-trunk arrangement with cylindrical symmetry. They could thus have very low thermal expansion coefficients analogous to needle cokes used in electrographite manufacture.

5.3   Electrical behaviour of carbons in general

5.3.0   Generalities

Polymeric materials containing only σ-bonds between carbon atoms in the sp$_3$ state are generally insulators with conductivities less than 10^{-15} Ω^{-1}cm^{-1}. When π-bonds associated with groups of carbon atoms in the sp$_2$ state are present, electrons are delocalized and are available as charge carriers. Polyvinylene ($-$CH$=$CH$-$)$_n$, for instance, has a conductivity of 10 Ω^{-1}cm^{-1}.

Organic polyaromatic hydrocarbon crystals are intrinsic semi-conductors with conductivities lying between 10^{-2} and 10^{-8} Ω^{-1}cm^{-1} and, characteristically, show an increase in conductivity as the test temperature is raised. This is associated with thermally activated transfer of electrons from molecule to molecule.

The extreme case of a polyaromatic material is that of graphite single crystals, the component sheets of which show two-dimensional metallic conduction with a conductivity greater than 3×10^4 Ω^{-1}cm^{-1} within the sheet and a negative temperature coefficient.

As we increase the proportion of conjugated carbon in the sp$_2$ state during pyrolysis, we shall change the material progressively from an insulator to a good conductor – a remarkable change of up to 19 orders of magnitude. The electrical properties will afford, therefore, a sensitive measure of the various stages of pyrolysis and provide information on the structure of the final carbon. The electronic changes observed in the growth and coalescence of polyaromatic sheets during the calcining and graphitization of coke will be compared and contrasted with those observed during the growth and coalescence of polyaromatic ribbons in polymeric carbons, which are our main interest here.

5.3.1   Electrical behaviour of crystalline polyaromatic hydrocarbons

The electrical behaviour of crystalline organic semiconductors has been reviewed previously by Inokuchi and Kamatsu (1961) and, latterly, by Gutmann and Lyons (1967).

The π-electrons in molecules of polyaromatic hydrocarbon crystals are delocalized within the molecule. As the number of sp_2 carbon atoms incorporated in the system is increased, the π-electrons become more and more delocalized from the parent carbon atom and the ionization potential decreases. However, molecule edges are screened with hydrogen atoms and the large separation between molecules offers an effective barrier to the easy transfer of electrons from one molecule to the next. In terms of band theory, the weak intermolecular attraction is associated with narrow bands with high effective masses and, consequently, low mobilities and conductivities (McLintock and Orr, 1973). The predicted band width is so narrow that the mean free path of the carriers is less than the intermolecular distance. Under these circumstances, only hopping and tunnelling mechanisms can operate.

Polyaromatic hydrocarbon crystals, therefore, act as semiconductors. As the test temperature is raised, the conductivity increases. A classical plot of $\log \sigma$ against T^{-1} normally yields a straight line:

$$\log \sigma(T) = \log \sigma_\infty - E_g/2kT$$

where E_g is the 'energy gap'. In many cases, kinks are observed in this plot, yielding apparently lower energy gaps at lower temperatures. This effect is attributed to the presence of impurities and defects.

Inokuchi (1951) found that both E_g and $\log \sigma_\infty$ are inversely proportional to the polyaromatic area. He explained this in terms of a model in which the polyaromatic molecules act as the plates of a condenser. However, variations in E_g can also be obtained by variations in humidity, in the case of molecules with polar adsorption sites, or solvent vapour pressure, in the case of polyaromatics. Thus the separation between molecules is also important.

Many have found (cf. Rosenberg *et al.*, 1968) that in all such organic materials, E_g is related simply to the limiting conductivity (σ_∞) by

$$\log \sigma_\infty = (2kT_0)^{-1} E_g + \log \sigma'_\infty$$

where T_0 is a 'characteristic temperature'.

Eley (1967) plots $\log \sigma_\infty$ against E_g for a wide range of organic systems. No explanation is given.

Gutmann (1967) has developed a model for organic semiconductors from which he calculates that below 1000 °C, hopping becomes negligible leaving tunnelling as the only possible mechanism for electron transfer at normal temperatures. In order to explain the observed sensitivity to changes in temperature, since tunnelling is not expected to be so temperature dependent, Gutmann proposes that neighbouring molecules must be favourably oriented to each other for tunnelling to occur. This transient reorientation will be thermally activated; it can involve either movement of the complete molecule or merely internal vibrations and rotations within a complex molecule.

5.3.2 Electrical behaviour of graphite crystals

A detailed description of the electrical behaviour of graphite approaching perfection is given by Spain (1973). With extensive sheets of graphite containing few defects, electrical conductivities greater than 3×10^4 $\Omega^{-1} cm^{-1}$ at room temperature in the plane of the sheet have been measured. The conductivity decreases as the test temperature is raised indicating that conduction is metallic and the scattering process mainly thermal. The highest filled valence band overlaps the lowest empty conduction band by ~36 meV. Even at temperatures approaching absolute zero there are empty valence states (holes) and filled conduction states (electrons). The valence band arises from the bonding (π) orbitals and the conduction band from the antibonding (π^*) orbitals. The conductivity perpendicular to the sheets is much lower, reflecting the weaker van der Waals bonding between layers; values of only ~1 $\Omega^{-1} cm^{-1}$ have been reported at room temperature. Raising the temperature increases the conductivity, presumably because thermal activation allows electrons to jump or tunnel from one sheet to the next. A value of 0.65 eV has been quoted (Ubbelohde, 1959) for the excitation energy. It is inferred that good conductivity is only possible in carbons if there is a continuous polyaromatic path through the material.

The conductivity of graphite is sensitive to defects, either those inflicted by irradiation or those grown-in during its formation. These defects can involve broken bonds in the case of interstitials and vacancies or unit c-axis dislocations (cf. Jenkins, 1973) which may either be isolated or grouped in array to form a tilt boundary. In such cases the interaction with electrons is strong and long range. The defects could involve no C–C rupture as in the case of simple bend and warp in the graphite sheets; in this case, the interaction with electrons is difficult to predict. As the population of any of these defects increases, the conductivity decreases, presumably because the mean free path of charge carriers is drastically lowered. The variation of conductivity with temperature changes from negative to positive indicating that the interference of lattice defects with electron movement is temperature dependent and the material is converted formally into a semiconductor.

Defects in crystalline intrinsic semiconductors, such as GaAs, increase the number of available states and affect the mobility only marginally. The mobility of charge carriers in graphite, on the other hand, is enormously reduced in the presence of defects. In graphite, this effect is exacerbated because the lattice can sustain an extraordinarily high stable defect population.

5.3.3   Electrical behaviour changes during the heat-treatment of cokes

The growth of conductivity in cokes has been studied in great detail because of the commercial interest in the manufacture of graphitized electrodes. Mrozowski (1952) first presented a plausible interpretation of the changes occurring in cokes from the temperature of formation at ~450 °C up to graphitizing temperatures using a band model usually applied to perfectly crystalline materials.

As hydrogen and excess low-molecular weight hydrocarbons are removed from the peripheries of the condensed ring systems, some of the σ-electrons from the peripheral carbon atoms are left unpaired (some, of course, become bonded). A π-electron will now jump from the π-band into the σ-state, forming a spin pair. This effectively removes an electron from the π-band and creates a hole in the filled band, leading to p-type conduction. A large number of holes are created which accounts for a great increase in conductivity from $10^{-5}\ \Omega^{-1} cm^{-1}$ to $10^{+2}\ \Omega^{-1} cm^{-1}$ between 500 and 1200 °C. Eventually, the number of holes becomes so great and the π-band so depleted of electrons that p- changes to n-type conduction at 900 °C. The energy gap between the π-band and the conduction band decreases as the conductivity increases, dropping from 0.62 eV to 0.03 eV.

The number of carriers decreases with the gradual elimination of spin pairs but, simultaneously, the free path increases with consequent growth in layer diameter. The result is a constant conductivity over the heat-treatment range 1200 to 1700 °C. Concurrently, over this range, the Fermi level rises and electrons start refilling the π-band. Eventually, at 1700 °C the conduction changes from n- to p-type. As the temperature is raised further, the number of carriers increases because the energy gap becomes negligible and a great number of electrons are raised thermally into the conduction band. The Hall coefficient, therefore, moves dramatically through a positive maximum at 2000 °C and thereafter falls steeply to become negative again above 2300 °C, as three-dimensional order is established in the graphite crystallites. Both the increase in mean free path and the number of carriers cause a further reduction in resistivity to 10^{-3} Ωcm. In well-graphitized material the number of electrons probably equals the number of holes but the greater mobility of electrons causes the Hall coefficient to remain slightly negative.

In recent years, Mrozowski (1971) has modified this simple picture. He now considers that the band gap disappears after calcining removes most of the hydrogen, typically, at 1200 °C. Overlap is now thought to be pronounced above 2000 °C.

5.3.4 Changes in electrical behaviour during the heat-treatment of polymeric carbons

When polymers are pyrolysed to form polymeric carbons, the same pattern as that for cokes is followed in that the pyrolysis regime between 400 and 1500 °C is characterized by an enormous drop in resistivity, this time from $\sim 10^{18}$ Ωcm to 3×10^{-3} Ωcm. Generally, the drop in room temperature resistivity with heat-treatment temperature (θ) follows a smooth relationship of the form:

$$\log_{10}[\log_{10}(\rho/\rho_\infty)] = -\theta/\theta_c + I$$

where, typically, $\rho_\infty = 3 \times 10^{-3}$ Ωcm

$$\theta_c = 300 \text{ K}$$
$$I = 3.6$$

Thus to bring ρ to within an order of magnitude of the limiting resistivity, ρ_∞, the pyrolysis temperature must rise, typically, to 1080 K. This is merely a useful rule-of-thumb and does not have any fundamental significance, that we are aware of.

The very early stages of pyrolysis are characterized by a marked time dependence (Bücker, 1973). On applying a field to a phenolic resin pyrolysed below 500 °C, for instance, the current decreases with time and a stationary value is reached only after hours or even days. This is caused by dielectric relaxations. Antonowicz et al. (1973) claim that a switching effect occurs in polyfurfuryl alcohol pyrolysed below 600 °C. The memory time decreases as the pyrolysis temperature is raised but, even at 600 °C, a sample could be switched between low and high resistance states.

Pyrolysis between 400 and 500 °C is also characterized by a rapid growth in the free radical concentration, as determined by electron spin resonance, to $\sim 4 \times 10^{20}$ g^{-1} in the case of PVDC, for instance (Blayden and Westcott, 1963). Further temperature increase to 700 °C is characterized by a sudden drop in this free radical concentration. This is interpreted as signifying that edge carbon atoms in polymeric networks are left with dangling bonds as foreign atoms are stripped off. These are satisfied by coalescence with neighbouring edge atoms to form wider ribbons as the temperature of pyrolysis is raised.

Yamaguchi (1964) first measured the electrical resistivity of glassy carbon subjected to heat-treatment at temperatures between 800 and 3200 °C; he remarked on the steep decrease observed between 800 and 1200 °C. Yamaguchi found that the Hall coefficient was much smaller in general than that of graphitizing carbon. Up to a heat-treatment temperature of 1000 °C, the Hall coefficient was positive; it became negative and passed through a minimum value at about 1400 °C. It finally became positive again at 1600 °C. Thereafter, the Hall coefficient rose slowly with further increase in heat-treatment temperature. This behaviour differs from that of graphitic carbons which show a sharp positive maximum at 2000 °C followed by a rapid decrease to negative values as graphitization proceeds. Yamaguchi (1964) found that the Hall coefficient of polymeric carbons was independent of the ambient temperature. Tsuzuku and Saito (1966) also showed it to be independent of magnetic field intensity over a wide range.

Yamaguchi concluded that the electrical behaviour of glassy carbon corresponds to that of graphitizing carbons heat-treated at low temperatures and so a two-carrier conduction mechanism is not applicable to glassy carbon; the Fermi level is very low even on annealing at 3200 °C.

Yamaguchi (1964) measured the thermo-electric power of glassy carbon annealed at 3000 °C at test temperatures up to 3000 °C. He observed that thermo-electric power decreases with increase of test temperature above 600 °C and changes sign to negative at 2000 °C. This indicates that glassy carbon becomes a partly two-carrier system above 600 °C and is electronically similar to graphitizing carbons above 2000 °C.

Tsuzuku and Saito (1966) measured the thermo-electric power of glassy carbon as a function of heat-treatment temperatures between 800 and 3000 °C at three ambient temperatures of −150, 150 and 500 °C. They showed that thermo-electric power was dependent on the test temperatures and, in general, higher values of thermo-electric power were obtained at higher test temperatures. The thermo-electric power was positive at a heat-treatment temperature of 1000 °C, became negative at 1300 °C and decreased to a minimum at 1500 °C. The value became positive again at 1700 °C and grew with further rises in heat-treatment temperature. These results were quite different from those of non-graphitizing carbon made from phenol–benzaldehyde, which exhibits partial graphitization (Loebner, 1955). In fact, Yamaguchi (1964) reported different results for electrical properties measured in a porous glassy carbon, in which he considered that partial graphitization took place. In these cases, electrical properties were more characteristic of graphitic behaviour.

Yamaguchi measured the magneto-resistance of glassy carbon and found that it showed negative values and decreased continuously with increase of heat-treatment temperature up to 3200 °C. It has been found by Mrozowski and Chaberski (1956) that graphitizing carbon shows a small negative magneto-resistance at a heat-treatment temperature of about 2000 °C, the negative value changing rapidly to positive and increasing with further increase of heat-treatment temperature. After studying the Hall coefficient of graphitizing carbons, Mrozowski and Chaberski (1956) suggested that negative magneto-resistance could be associated with holes and positive magneto-resistance with electrons. Although no satisfactory explanation of the mechanisms of the negative magneto-resistance are given, the negative values of glassy carbon throughout the whole range of heat-treatment are characteristic of extreme non-graphitizability.

Room temperature electrical resistivity of glassy carbon derived from phenolic resin was measured by Kawamura (1971) using a potentiometric method; fig. 24 shows its variation with heat-treatment temperature. The resistivity decreases rapidly up to a heat-treatment temperature of 900 °C and thereafter decreases gradually reaching a constant value at 1500 °C. It falls slightly at a heat-treatment temperature of 2000 °C, as was reported by Yamaguchi (1964), and rises at 2250 °C. The resistivity decreases gradually after heat-treatment above 2500 °C. The characteristic features are the steep decrease of the resistivity during pyrolysis and the surprisingly low resistivity of the resulting carbon, comparable to that of well-graphitized material with crystallite diameter greater than 1000 Å.

Since X-ray diffraction studies show that the resulting carbon has very small crystallites which are randomly oriented, the mean free path should be very small. This suggests that the carrier density is high. The small values of the Hall constant measured by Yamaguchi also indicate a high carrier concentration.

The carrier density of partially carbonized organic substances is closely related to the proportion of foreign atoms. The only foreign atom above 500 °C is hydrogen. It is therefore important to investigate the relationship between electrical resistivity and the amount of hydrogen retained. The hydrogen/carbon atomic ratio decreases significantly be-

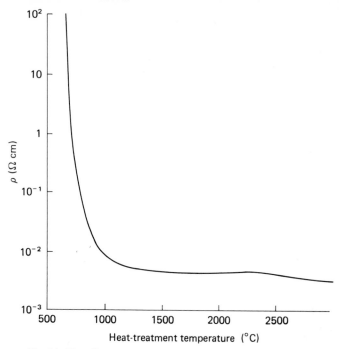

Fig. 24. The effect of heat-treatment temperature on the electrical resistivity (ρ) of a polymeric carbon.

tween 500 and 650 °C. Figure 25 reveals that the logarithm of the room temperature electrical resistivity is proportional to the hydrogen/carbon atomic ratio at heat-treatment temperatures between 650 and 1000 °C. The curve shows a marked kink between 675 and 700 °C. The electrical resistivity decreases more rapidly with decrease of hydrogen/carbon ratio at heat-treatment temperatures below 700 °C than above 700 °C. This is qualitatively explained by assuming that the removal of hydrogen leaves free radicals at the periphery of the condensed aromatic molecules. These free radicals can produce mobile carriers according to the type of reaction proposed by Pohl (1962) in which an edge carbon atom attracts electrons from a carbon atom in the interior of the sheet to produce a negatively charged edge ion and a mobile interior hole. This results in an increase in the number of mobile carriers and leads to a steep decrease in electrical resistivity. As the distance between the condensed aromatic molecules becomes smaller with the release of internal strain energy, the free radicals at the edges form cross-links between the aromatic molecules. The formation of cross-links decreases the potential barrier between aromatic layers, but does not increase the number of mobile carriers. The recombination of mobile holes and

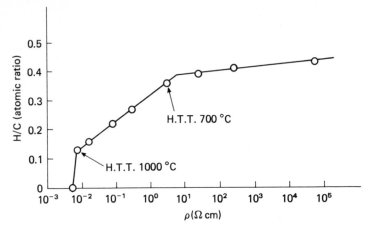

Fig. 25. Variation of the electrical resistivity (ρ) of a polymeric carbon with the hydrogen/carbon ratio.

electrons derived from what were previously bound edge ions reduces the population of mobile carriers.

The slower decrease of electrical resistivity with reduction in hydrogen/ carbon ratio observed at heat-treatment temperatures above 700 °C should be related to the decrease in mobile carriers with the formation of intermolecular cross-links.

Figure 26 shows the room temperature electrical resistivity on a logarithmic scale as a function of volume shrinkage and bulk density. Both curves reveal that there is a marked change in electrical conduction at a heat-treatment temperature of about 700 °C. In general, the electrical resistivity of a carbon depends considerably on its bulk density. The electrical resistivity of porous carbon discs of density 1.2 gcm³ after being heat-treated at 1000 °C shows a similar behaviour to that of the non-porous carbon as a function of heat-treatment temperature, but its value is one order of magnitude higher.

To compare the ultimate resistivity of isotropic polymeric carbon with that of graphite single crystals, it is necessary to make a correction for density. It is very easy for us to introduce porosity into our glassy carbon and thus to alter its density by adjusting the conditions prevailing during the sintering of the original resin. A range of materials so formed provides us with a means of relating resistivity to bulk density, varying from 0.8 to 1.5 gcm⁻³. In fig. 27, we plot the conductivity directly against the density. It is possible to discern two linear regions of different slope. At densities below 1.15 gcm⁻³, optical microscopy shows the material to consist of discrete grains. Above 1.15 gcm⁻³, the pores are separated and surrounded by an otherwise continuous solid matrix.

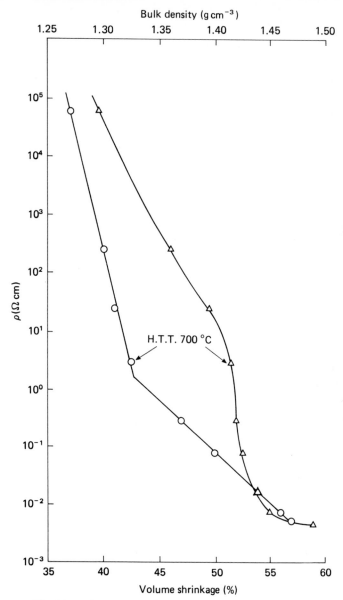

Fig. 26. Variation of the electrical resistivity (ρ) of a polymeric carbon with its bulk density. ○, bulk density; △, volume shrinkage.

The straight line characteristic of the latter region can be extrapolated to a resistivity value of $10^{-3}\ \Omega$cm at a density equivalent to that of graphite single crystals (2.26 gcm^{-3}). This is a factor of 20 greater than

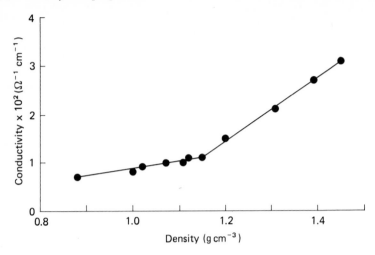

Fig. 27. Electrical conductivity of glassy carbons annealed at 2700 °C as a function of density.

that calculated to be the resistivity of a randomly oriented mass of perfect graphite crystals, thus indicating by how much the folds and narrowness of polyaromatic ribbons in a polymeric carbon affect their electrical conductivity.

The resistivity of polymeric carbon fibres with a range of degrees of preferred orientation has been measured (Ezekiel, 1970; Bacon and Schalamon, 1969). The axial resistivity ρ_a was shown to vary inversely as the axial modulus E_a, such that $\rho_a \times E_a \approx 23 \times 10^6 \ \Omega \mathrm{Nm}^{-1}$. Clearly the axial resistivity is a function of the degree of alignment of the component microfibrils in the same way as the elastic compliance. With perfect orientation ρ_a approaches $3 \times 10^{-4} \ \Omega \mathrm{cm}$, which is an order of magnitude greater than the resistivity of perfect graphite sheets.

5.3.5   Polymeric carbon as a semiconductor

Many carbonaceous materials show increases in conductivity with increase in test temperature and so are termed semiconductors. Crystalline polyaromatic hydrocarbons are intrinsic semiconductors. Defective graphites are semiconducting by virtue of barriers raised against charge carrier transport by the presence of stable defects. The causes of semiconduction in polymeric carbon is the subject of much current research.

Tsuzuku and Saito (1966) first reported substantial increases in conductivity with test temperature, using glassy carbon heat-treated at

800 °C. The plot of resistivity in logarithmic scale versus the inverse of the absolute temperature was found to give a non-linear curve. Kawamura (1971) has extended these measurements to lower heat-treatment temperatures using our glassy carbon.

Figure 28 shows examples of the variation of electrical resistivity on a logarithmic scale as a function of the reciprocal of ambient temperature in our samples of pyrolysed phenolic resin from room temperature to 400 °C in a dry nitrogen atmosphere. The electrical behaviour is sensitive as to whether the ambient gas is air, nitrogen or argon at low heat-treatment temperature. The resistivity clearly decreases with rise in test temperature and the greatest effect is obtained in samples heat-treated at lower temperatures.

A characteristic feature is a marked kink at a test temperature of about 250 °C throughout the whole range of heat-treatment temperatures up to 2700 °C. Below 750 °C heat-treatments, the curves show several kinks, but only the kink at a test temperature of 250 °C remains in samples treated above 750 °C.

The electrical resistivity (ρ_T) of glassy carbon heat-treated above 750 °C is clearly expressed adequately by

$$\rho_T = \rho_\infty e^{E_g/2kT} \quad \text{above 250 °C}$$

and

$$\rho_T = \rho'_\infty e^{E'_g/2kT} \quad \text{below 250 °C.}$$

Although it may not be appropriate to apply band theory to such a non-crystalline material, the formulae suggest that the pyrolysed resin behaves as an intrinsic semiconductor with 'impurity levels'. The intrinsic energy gap (E_g) can be calculated from the slope in fig. 26 at test temperatures above 250 °C and the energy gap due to 'impurities' (E'_g) from the slope below 250 °C. Both E_g and E'_g are listed in table 7 as a function of heat-treatment temperatures.

TABLE 7   *The variation of energy gaps with heat-treatment temperature*

Heat-treatment temperature (°C)	E_g intrinsic energy gap (eV)	E'_g 'impurity' energy gap (eV)
650	0.81	0.61
700	0.34	0.26
750	0.18	0.15
900	0.12	0.059
1000	0.085	0.039
1500	0.035	0.0064
2000	0.034	0.0096
2500	0.034	0.012
2700	0.033	0.016

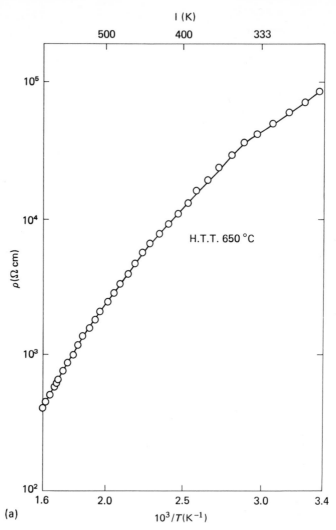

Fig. 28(*a*) and (*b*). Variation of the resistivity (*ρ*) with temperature of measurement in a phenolic resin carbonized and annealed progressively from 650 to 2500 °C.

The 'intrinsic' energy gap decreases continuously with increase in heat-treatment temperature. There is some resemblance with intrinsic organic semiconductors in that the 'intrinsic' energy gap varies linearly with the logarithm of the room-temperature resistivity (ρ_0) such that

$$\log_{10} \rho_0 = 9.1\, E_g - 2.6$$

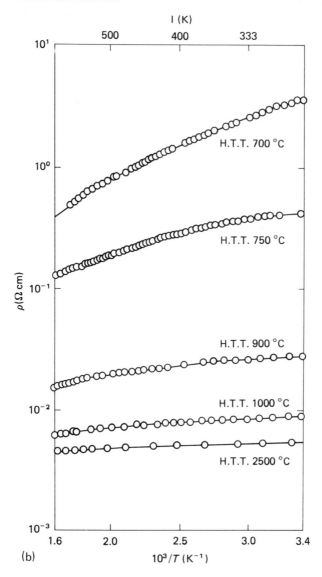

(b)

The 'impurity' energy gap also follows $\log_{10} \rho_0$ up to $\theta = 1500\ ^\circ C$ in a similar manner:

$$\log_{10} \rho_0 = 13.8\, E'_g - 2.4$$

Above 1500 °C, however, the 'impurity' gap ceases to follow the resistivity; it now increases.

Both the intrinsic energy gap and the electrical resistivity are very dependent on the hydrogen content. It is presumed that E_g is closely related to the potential barrier between condensed aromatic molecules, which is a function of the amount of hydrogen retained in the system.

Another important factor determining the energy gaps is the surface area of the condensed aromatic molecules. Inokuchi (1951) and Schuyer and Van Krevelen (1955) found that the intrinsic energy gap is proportional to the reciprocal of the surface area of the condensed aromatic molecules in polynuclear aromatic compounds and carbonized coals, i.e.

$$E_g = 100/S - 0.25$$

where E_g is intrinsic energy gap in eV and S is the surface area of the individual condensed aromatic molecule in 10^{-2} nm^2 units.

Figure 29 illustrates the variation of the surface area of a condensed aromatic molecule and the square root of the surface area as a function of heat-treatment temperature. The calculated square root of the surface area of the condensed aromatic molecule is about half the crystallite diameter measured from the line broadening of X-ray diffraction profiles. Both the surface area and its square root increase considerably

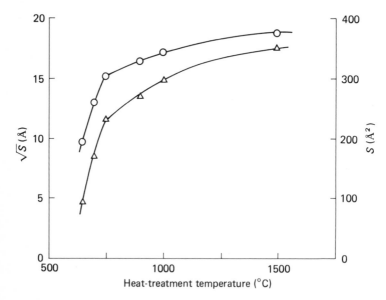

Fig. 29. Variation of aromatic surface area (S), calculated from a condenser model, with heat-treatment temperature of a polymeric carbon. ○, $\sqrt{S}$; △, S.

up to a heat-treatment temperature of 750 °C and thereafter increase much more slowly. Between 1500 and 2700 °C, the calculated surface area of the aromatic molecules remains constant. X-ray diffraction studies show that the apparent crystallite diameter (L_a) increases continuously over this temperature range. This suggests that the physical meaning of the surface area calculated from the intrinsic energy gap is different from that calculated from the line broadening of X-ray diffraction profiles. It may be that the surface area calculated from the intrinsic energy gap provides the domain size of a graphitic layer surrounded by electronic disorder associated with vacancies, interstitials or non-trigonal carbon atoms, while the crystallite diameter, L_a, is a distance between more widely dispersed boundaries or high concentrations of defects such as high-angle misfit between graphitic crystallites. These physical meanings may be coincident at low heat-treatment temperatures.

An interesting and important characteristic in fig. 28 is the kink observed at a test temperature of about 250 °C in all samples heat-treated up to 2700 °C. Waters (1961) found a similar kink in partially carbonized coals and suggested that the secondary current carriers are due to free radicals or unpaired electrons. It is reasonable to assume the presence of free radicals in carbonaceous materials carbonized at low temperatures, but these free radicals generally become paired off at 700 °C. Yet glassy carbon shows the kink even after annealing at 2700 °C. Tsuzuku and Saito (1966) also observed a similar kink in commercial glassy carbons heat-treated to 3000 °C. It may be necessary to propose a different mechanism for secondary current carriers in glassy carbon.

Waters also proposed a mechanism for secondary current carriers, based on the assumption that carbonized coal has aromatic molecules of varying size. Provided the diameter of the crystallites varied widely, the secondary current carriers could be the more loosely held π-electrons associated with larger aromatic molecules. Since glassy carbon is a non-graphitizing carbon and its X-ray diffraction peaks are highly asymmetric even after annealing at high temperature, it seems reasonable to assume that glassy carbon is also made up of condensed aromatic molecules of widely varying size. However, if one assumes that the 'impurity' level is due to loosely held π electrons in larger graphitic crystallites, the crystallite diameter responsible for the 'impurity level' must decrease with increase of heat-treatment temperature above 1500 °C, because the energy gap due to 'impurities' then increases; this is difficult to accept.

In our structural model, glassy carbon is composed of long aromatic ribbon molecules with no clearly defined crystallite boundaries. If these ribbons are perfect graphite planes, the energy gap will be zero (Wallace, 1947). However, the ribbons are highly strained; energy gaps could be attributed to a distribution of strain along the aromatic ribbons, the intrinsic energy gap being related to a major distortion in the ribbon network. The presence of 'impurity levels' suggests that there are at least two different types of straining in the network, especially when one notices that the 'impurity level' increases with heat-treatment above 1500 °C, while the intrinsic energy gap remains fairly constant. The increase in the 'impurity' energy gap may be ascribed to an increase in local strain in the network. It is tempting to associate this with strains induced by the anisotropic contraction of crystallites on cooling from high annealing temperatures. The higher the firing temperature, the greater the internal strain and so the greater will be the associated energy gap.

As was previously discussed, the 'impurity levels' may also be ascribed to the presence of free radicals created during carbonization. It is not clear, however, why carbonaceous materials should have a considerable concentration of free radicals after annealing at high temperature. When carbonization is complete, at about 1250 °C, these free radicals should disappear. The concentration of free radicals could be further increased by breaking carbon–carbon bonds to attain ribbon perfection. The 'impurity levels' observed after annealing above 1500 °C could be due to these newly created free radicals, which can also trap π electrons to form negative ions at the edges of the ribbons.

The behaviour of the Hall coefficient and thermoelectric power of glassy carbon as a function of heat-treatment temperature shows the material to be a p-type semiconductor except in the region of heat-treatment temperature between 1300 and 1600 °C. In terms of the free radical theory, the transition from p to n at 1300 °C would be due to the decrease of electron traps as intermolecular cross-linking proceeds, while the transition back to p, at 1600 °C would be due to the trapping of electrons at newly created free radicals. The free radical interpretation also agrees with ESR studies carried out by Tsuzuku and Saito (1970), who observed that the spin centre concentration in glassy carbon is greatly diminished by heat-treatment to 1300 °C. Heat-treatment to higher temperatures causes an increase in spin centres. They claim that spin centres are partially localized even in glassy carbon heat-treated at 3000 °C.

Subsequent to our work on pyrolysed phenolic resin rods, Bücker (1973) has measured the electrical behaviour of phenolic resin films, originally 2 mm in thickness, pyrolysed up to 1200 °C. He finds that the plot of log σ against T^{-1} is not resolveable into linear portions between 200 and 500 K, but is a continuous curve. A plot of log σ against $T^{-1/4}$ yields straight lines of form

$$\log \sigma = \log \sigma_\infty - BT^{-1/4} \text{ for constant } \theta$$

where B decreases with increase in θ in a similar manner to our calculated band gaps E_g, E_g'. log σ_∞ decreases rapidly as the temperature of pyrolysis is raised, from

$10^{28} \ \Omega^{-1}\text{cm}^{-1}$ for $\theta = 400$ °C to $3 \times 10^2 \ \Omega^{-1}\text{cm}^{-1}$ for $\theta = 800$ °C.

Thereafter, it remains constant.

Like us, Bücker finds that:

$$\log(\sigma/\sigma_\infty) = -C \, \mathrm{d}(\log \sigma)/\mathrm{d}(kT)^{-1} \text{ for constant } T.$$

Thus the 'band gap', given by $\mathrm{d}(\log \sigma)/\mathrm{d}(kT)^{-1}$, varies linearly with the logarithm of the conductivity. Incidentally, C can be evaluated from the last two equations – it becomes simply $C = (4kT)^{-1}$, if σ_∞ can be identified with σ_∞'. For heat-treatment between 600 and 1200 °C, σ_∞' is constant and approximately equal to σ_∞. For pyrolyses below 600 °C, σ_∞' changes to $\sim 1 \ \Omega^{-1}\text{cm}^{-1}$.

Because of the conformation with a $T^{-1/4}$ law, Bücker suggests that polymeric carbon should be included in a class of 'amorphous' materials made by deposition from the vapour phase, including films of Ge, Si, GaAs, GaSb and carbon. Adkins *et al.* (1970) and McLintock and Orr (1973) had previously shown that carbon films produced by arc-evaporation follow the $T^{-1/4}$ law. These materials have a great number of unsatisfied bonds and are without long-range order.

The electrical behaviour of such materials has been described by a general theory of electrical conduction in 'amorphous' materials. In this theory, the concepts of allowed energy bands and forbidden gaps used for crystalline semiconductors are retained. However, the edges of the bands are ill-defined and a tail of electron states extends into the forbidden gap. Near band edges the states are localized in the tail of states extending into the gap. At energies corresponding to localized states, conduction takes place by diffusion, hopping or tunnelling, with mobilities much lower than those of free carriers. This leads to a 'mobility gap'. McLintock and Orr (1973) suggest that the low carrier mobilities in disordered carbons are consistent with conduction via such localized states.

Mott and Davies (1971) consider a specific conduction mechanism in such materials in which the current depends on the mobility of carriers hopping or tunnelling between localized states close to the Fermi energy. This is formally analogous to impurity conduction in heavily doped single crystals. To obtain a $T^{-1/4}$ law, it is necessary to conceive of a curious low temperature mechanism called 'variable range hopping' in which the charge carrier has a choice of centres to hop to, which may not necessarily be adjacent. With more centres from which to choose, the energy difference can be smaller.

Expressed mathematically, only electrons with energies in the range of order kT at the Fermi energy (E_f) can take part. The number of such electrons is $N(E_f)kT$ where $N(E_f)$ is the density of states near the Fermi level. The exponent of the jump frequency is given by

$$-2\alpha R - W/kT$$

where R is the distance covered in each hop, W is the mean activation energy for hopping and is inversely proportional to $N(E_f)$, and α describes how the wave function decreases with distance from a centre, assuming spherical symmetry.

Substituting for W and R, which become functions of T for variable range hopping, the exponent of the jump frequency becomes $-BT^{-1/4}$, where

$$B^4 = 18\alpha^3/kN(E_f)$$

Thus B is dependent on α and $N(E_f)$. The structural significance of these parameters is not yet elucidated for the case of carbons.

It is surprising that such a mechanism, which Mott insists can only predominate at temperatures well below 200 °K, should extend to test temperatures in excess of 500 °K in the case of polymeric carbons studied by Bücker.

McLintock and Orr (1973) have adapted Gutmann's model for organic semiconductors (Gutmann, 1967) (discussed in section 5.3.1) to try to explain the conducting process in disordered carbons. Disordered carbons contain components which are loosely held in the molecular framework. These are free to oscillate or rotate. The poly-aromatic layers are large enough to permit internal atomic movements or electronic polarizations that are a feature of Gutmann's realignment theory. The rotational 'clicking' model of McLintock and Orr (1973), which they use to explain the removal of spin centres with heat treatment, can be regarded as a special reversible case of such realignment. It is not clear how Mott's theory for low temperature conduction and Gutmann's mechanical vibration model can be brought together into a convincing explanation for semiconduction in polymeric carbons.

It should be emphasized that all these models predict that conductivity should vanish at zero degrees Kelvin. However, measurements on PAN pyrolysed at 950 °C by Helberg and Wartenberg (1970) show that polymeric carbon still has a conductivity greater than $10^2 \, \Omega^{-1} \text{cm}^{-1}$ at a temperature of 1.7 K and probably possesses an appreciable finite value, like graphite materials, as $T \to 0$ K. It is inferred that at a surprisingly early stage in the pyrolysis, a tenuous, defective and strained conduction path of polyaromatic ribbons is established. Examination of Bücker's results would put the pyrolysis temperature at which this occurs as low as 600 °C. Heat-treatment above 600 °C allows augmentation of the width of the ribbons, making the path less tenuous, and progressively removes defects and strains. However, even after extensive annealing at 3000 °C, major folds and the close proximity of ribbon edges prevent the glassy carbon from possessing the high mobility of charge carriers and n-type conduction which are characteristic of well-graphitized materials.

Herinckx (1973) has made measurements on carbon fibres showing that the conductivity increases with θ and the degree of preferred orientation. He also adopts a 'ribbon' model for polymeric carbon and attempts to use the language of amorphous semiconductor theory. Localized states are attributed to the large number of defects at the edges of ribbons. He finds that there is a constant density and mobility of free carriers (mainly holes) at low temperatures. This he also attributes to the states being located in a small region near the limit of the mobility gap in the valence band. Calculations show a close correlation between the mobility and structural defects in the fibre. He emphasizes that the greatest influence on electrical conductivity is the density of structural defects such as local curvatures and ribbon edges which scatter the free charges. This effect is much more important than the increase in carrier density resulting from edge atoms acting as acceptors.

5.3.6   Piezo-resistivity

We have argued in the previous sections that internal strains are a possible cause of the high resistivity and the presence of 'band-gaps' in polymeric carbons. It is logical to ask if externally applied stresses would produce similar effects on electrical behaviour.

The sensitivity of electrical resistance in carbon to applied stress is well known and is made use of in microphones. The usual material used is heat-treated anthracite. In Swansea, Takezawa and Jenkins (1974) have measured the fractional change in electrical resistance on subjecting rods of carbons derived from phenolic resin to tensile loads. Material heat-treated to 700 °C was extremely sensitive ($3 \times 10^{-11} \, \text{cm}^2 \text{dyne}$). Even when the change in shape was taken into account to measure the true change in resistivity (piezo-resistivity), the sensitivity was about 2×10^{-11} cm^{2}dyne^{-1}. Further heat-treatment to 1200 °C reduced this effectively to zero. It is postulated that the piezo-resistivity is directly related to the intrinsic band-gap of the polymeric semiconductors.

When carbon fibres are pulled (Conor and Owston, 1969; Owston, 1970), most show an increase in resistivity. However, those with the highest modulus and electrical conductivity show a decrease, presumably because they are more sensitive to increases of alignment of component microfibrils. Berg *et al.* (1971) have shown that a torsional strain introduced a non-reversible change in resistivity in carbon fibres. By way of explanation, Owston (1970) suggests that fibres have a relatively 'loose' microfibrillar structure which is opened out either by tensile or torsional stress. This implies that the overall resistivity is greatly influenced by the contact resistance between neighbouring microfibrils.

5.3.7 Magnetic susceptibility

Graphite possesses a remarkably high diamagnetic susceptibility. Many years ago, Krishnan and Ganguli (1939) reported that single crystals of natural graphite were highly anisotropic with respect to diamagnetic susceptibility. A value of -21.5×10^{-6} cgs g^{-1} was measured along the *c*-axis and only -0.5×10^{-6} cgs g^{-1} along an *a*-axis. Krishnan's postulate, by way of explanation, that the π-electron system behaves as a two-dimensional gas has been developed by Mrozowski (1952) and Pacault *et al.* (1960). Pacault considers that the electron gas behaviour can be attributed to the presence of vacancies in the π-electron energy distribution of each layer, rather than the total number of electrons. The diamagnetism is then determined by the concentration of charge carriers and their effective mass. The diamagnetic susceptibility of such a system increases exponentially with decrease in the test temperature, approaching a characteristic limiting value at absolute zero.

The diamagnetic susceptibility is highly dependent on the mean aromatic surface area, i.e. on the number of carbon atoms contributing to a graphite layer. There is a simple relationship between the susceptibility and the crystallite size (L_a) as determined by X-ray techniques. Kiive and Mrozowski (1959) studied the carbonization and graphitization process by measuring diamagnetic susceptibility, observing a particularly large increase as L_a increased from 75 to 150 Å. Above 150 Å the susceptibility remained constant at a value lower than that predicted from the average of the single crystal values.

Much work has been done on the diamagnetism of cokes, but little information is available on polymeric carbons. The initial stages of pyrolysis in all carbons are characterized by low values of the magnetic susceptibility with both paramagnetic and diamagnetic components. Paramagnetism is most marked in polymeric chars pyrolysed at 550 °C (Blayden and Westcott, 1961) and is associated with the build-up of a high population of unpaired electron spins. This is associated with a positive dependence of the susceptibility on test temperature.

Heat-treatment to graphitizing temperatures causes a decrease in the paramagnetic component and an increase in the diamagnetic part. Heat-treatment of PVC graphitic coke from 1000 to 2000 °C (Blayden and Westcott, 1961) results in a steep change in magnetic susceptibility from zero to -6.0×10^{-6} cgs g^{-1}. This remains constant with heat-treatment to higher temperatures. On the other hand, the magnetic susceptibility of PVDC char (polymeric) rises much less rapidly over the same heat-treatment range and even at 3000 °C reaches a value of only -4.7×10^{-6} cgs g^{-1}. For graphitic coke, the temperature dependence of the magnetic susceptibility changes sign from positive to negative at around 1300 °C. The same change occurs in polymeric char at 1500 °C. In both cokes and chars, these transitions are associated with the growth of the Landau diamagnetism, which is characteristic of perfect sheets of graphite. Fischbach (1967) measured the magnetic susceptibility of Japanese glassy carbon at room temperature as a function of magnetic field strength, heat-treatment temperature and high temperature tensile deformation. He observed that glassy carbon shows diamagnetic susceptibility which is independent of magnetic field strength. The diamagnetic susceptibility of glassy carbon increased with increase in the X-ray diameter and did not reach a constant value. Fischbach compared these results with those of graphitizing carbons, finding that the diamagnetic susceptibility of glassy carbon increases more rapidly with the apparent crystallite diameter than that of graphitizing carbons. For example, the diamagnetic susceptibility of GC-30 with an X-ray diameter of 65 Å corresponded to that of pyrolytic carbon with an apparent crystallite diameter of 130 Å. This result suggests that the significance of the apparent crystallite diameter is different in different classes of carbon materials.

He observed that tensile deformation causes a pronounced change in the anisotropy of susceptibility. Undeformed GC-20 samples had anisotropy ratios of 1.02, while GC-30 samples had ratios of 1.04. However, a GC-20 sample elongated 24% at high temperatures had an anisotropy ratio of 1.82, while a sample elongated about 15% had a ratio of 1.5. In each case, the low susceptibility direction was parallel to the tensile axis. Thus, it is clear that large tensile deformations introduced pronounced anisotropy in an initially isotropic material.

The anisotropy in susceptibility induced by tensile deformation was stronger in glassy carbon than in other isotropic polycrystalline carbons. This supports the structural model of glassy carbon in which tangled chains of graphite-like ribbons are linked edge-to-edge, because tensile deformation can then more easily align the chains parallel to the tensile axis at high temperatures.

6 Mechanical behaviour

6.0 Generalities

The mechanical properties of polymeric carbons are directly related to their structure, consisting of a network of graphitic ribbons with three types of bonding therein. Between parallel ribbons spaced at about 3.4 Å only weak van der Waals forces exist and these can be neglected. Within the ribbons, because of its extremely stiff bonding characteristic, the C–C bond in graphite is the stiffest component of the system. However, this is only felt with the high preferred orientation of some carbon fibres. For a random configuration of ribbons the response to an applied stress is related, in the main, to the ease of rotating ribbon segments with respect to their environment, and is resisted by edge-to-edge carbon bonding.

TABLE 8   *Mechanical properties of glassy carbons: the variations depend on the quality and heat-treatment temperature*

Compressive strength	100–700 GN m^{-2}
Tensile strength	50–200 GN m^{-2}
Flexural strength	60–140 GN m^{-2}
Shore hardness	80–120 GN m^{-2}
Charpy impact value	2–4 kg cm^{-1}

At graphitization temperatures, the edge-to-edge bonding can be broken at the behest of an applied stress and a preferred orientation of the ribbons is imposed in the direction of application.

A list of mechanical properties of a range of isotropic glassy carbons is given in table 8, the variations depend on the quality and the heat-treatment temperature.

A comparison between various carbons is given in table 9. It should be noted that stiffness and strength of polymeric carbons are greater than those of graphitic aggregates and, with high preferred orientation of component ribbons, approach the ultimate values observed in graphite whiskers.

For further reading, the reader is referred to Jenkins (1973).

TABLE 9   *Comparison of mechanical properties of various carbons*

	Tensile strength (MNm^{-2})	Stiffness (GNm^{-2})	Density (gcm^{-2})
A typical fine-grained isotropic graphite	50	15	1.9
A typical glassy carbon	200	28	1.45
Glassy carbon fibre	1800	40	1.5
PAN carbon fibre type I	3000	260	1.75
PAN carbon fibre type II	2000	400	1.95
Carbon-fibre-reinforced carbon	700	120	1.5
Graphite whiskers	>14000	1000	2.2

6.1   Elastic behaviour of graphite single crystals and polycrystalline material

The elastic deformation of a single crystal under a set of applied stresses may be described in terms of two alternative sets of elastic constants. We may write either

$$\text{Strain component} = S_{ij} \times \text{stress component},$$

where S_{ij} are called the compliances, or else

$$\text{Stress component} = C_{ij} \times \text{strain component},$$

where C_{ij} are the stiffnesses. For a hexagonal crystal of graphite, the following five compliances are chosen as independent constants:

S_{11} which is the inverse of Young's modulus along an a-axis;
S_{33} which is the inverse of Young's modulus along the c-axis;
S_{44} which is the inverse of the shear modulus governing the response to shear on a basal plane by a shear stress acting in the plane;
S_{12} which governs Poisson's ratio between principal strains in the graphite sheet;
S_{13} which governs Poisson's ratio between strains along a c-axis and along an a-axis;
The corresponding independent stiffnesses are C_{11}, C_{33}, C_{44}, C_{21}, C_{13}.

S_{11} is close to 10^{-12} m²N⁻¹, an extremely low value which reflects the stiff response to stretching a sheet of sp_2 carbon atoms. S_{33} is 27.5×10^{-12} m²N⁻¹ (cf. Blackslee *et al.*, 1970) – a much higher value which reflects the ease of pulling the graphite sheets apart. S_{44} is a variable quantity which depends on the interlamellar cohesion. Near-perfect graphite exhibits a high value of 5000×10^{-12} mN⁻¹ (Jenkins and Jouquet, 1968). On irradiation with neutrons this drops rapidly for small doses to 250×10^{-12} m^{2}N^{-1}. The other constants are rather indeterminate and, anyway, do not alter our predictions for poly-crystalline materials overmuch.

We are left with a picture of a material consisting of relatively in-extensible sheets which can be sheared over each other with ease, down to very low temperatures. Bending of the sheets is easy and invariably involves only shear strain between the sheets. Intracrystalline plastic deformation is easy, by glide of dislocations, between the sheets. These dislocations, although visible in electron micrographs of very perfect single crystals, are very diffuse. They are invariably split into partials between 800 and 1500 Å apart, the core width of each partial being greater than 50 Å. Plastic distortion of the sheets themselves can only take place at temperatures in excess of 2500 °C, when vacancies begin to move appreciably.

L_a in polymeric carbons is never greater than 100 Å even when the carbon is fully annealed. Thus, dislocation glide is hardly appropriate to describe deformation in such material.

Polycrystalline graphite, such as electrographite, has mechanical properties closely related to those of the component crystals. After graphitization it exhibits a low modulus and high static damping. Both are dependent on the average crystal dimensions, i.e. on the area of free parallel graphitic sheets. Irradiation increases the modulus of elasticity by a factor of 2.5 in the case of a typical electrographite, as opposed to 20 in a single crystal, indicating the extent to which sheet glide contributes to the stiffness.

6.2   Elastic response of polymeric carbons

Fischbach (1967) reported that dynamic Young's modulus measure-ments at room temperature on as-received materials gave values of 28 GN m^{-2} for GC-20 and 26 GN m^{-2} for GC-30. French glassy carbon V10, heat-treated at 1000 °C, has a modulus of 29 GN m^{-2}. Similar values have been reported by Taylor and Kline (1967). These results indicate a consistent decrease in modulus with increase of heat-treatment temperature.

These values are high for an isotropic carbon of low density. This is made clear in fig. 30 which compares the variation of the Young's modulus with density for isotropic pocographites, isotropic pyrocarbons and isotropic glassy carbon. The superior stiffness of the polymeric carbon is attributed to the strong boundary restraint at the edges of the component ribbons and strong interlamellar bonding at heat-treat-ment temperatures up to 2000 °C (Jenkins, 1973). The boundary

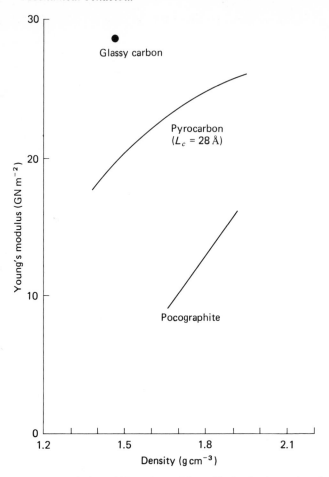

Fig. 30. Variation of Young's modulus with density for various isotropic carbons, illustrating the effect of boundary restraint.

restraint is less effective for the more extensive polyaromatic layers present in the pyrocarbons and pocographites.

Some attempts have been made to relate the elastic behaviour of polymeric carbons to the elastic constants of component crystallites, assuming these correspond with those of single crystals. Ruland (1969*a*) has undertaken this exercise on fibres and shows the inapplicability of either the uniform stress or uniform strain model. He adopts the standard formula for polymer fibres for the case of uniform stress:

$$E_c^{-1} = Q_z (S_{11} - 2S_{13} + S_{33} - S_{44}) + R_z (2S_{33} + S_4)$$

where E_c is the Young's modulus along the fibre axis for pore-free material and

$$R_z = \frac{\int \sin^3 \phi\, g(\phi)\, d\phi}{\int \sin \phi\, g(\phi)\, d\phi} \qquad Q_z = \frac{\int \sin^5 \phi\, g(\phi)\, d\phi}{\int \sin \phi\, g(\phi)\, d\phi}$$

where $g(\phi)$ = angular distribution of layer normals determined by X-ray scattering as defined in section 4.1 on structure.

For uniform strain the stiffness along the fibre axis is given by:

$$Q_z\,(C_{11} - 2C_{13} + C_{33} - C_{44}) + R_z\,(2C_{13} - 2C_{33} + C_{44}) + C_{44}$$

The predictions for Young's modulus along the fibre axis based on uniform strain and uniform stress theories are plotted in fig. 31 against the half-width, which is a measure of the degree of preferred orientation. Zero half-width signifies perfect axial orientation. The actual values collated by Diefendorf (1972) are presented for comparison. Quite arbitrarily, past workers have assumed that the polycrystalline modulus is the arithmetic mean of the two extreme models. It would seem from the inscribed curve that the geometric mean is much closer. We have no explanation for this.

Ruland rejects the mode for uniform strain because 'in a static stress–

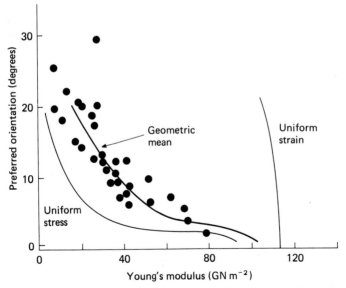

Fig. 31. Variation of axial Young's modulus of carbon fibres with preferred orientation compared with the theoretical extremes of uniform stress and uniform strain and their geometric mean.

strain experiment only the components of compliance can be measured directly'. He thus adopts the model:

$$E_c = \frac{\int g(\phi) \sin \phi / S(\phi)}{\int g(\phi) \sin \phi \, d\phi} \, d\phi$$

where

$$S(\phi) = \sin^4 \phi (S_{11} - 2S_{13} + S_{33} - S_{44}) + \sin^2 \phi (2S_{13} - 2S_{33} + S_{44}) + S_{33}$$

This is a much more complex relationship which has to be evaluated numerically for each of S_{11} and $g(\phi)$.

Using the single crystal values:

$S_{11} = 0.985 \times 10^{-12} \text{ m}^2 \text{N}^{-1}$
$S_{13} = -0.63 \times 10^{-12} \text{ m}^2 \text{N}^{-1}$
$S_{33} = 27.8 \times 10^{12} \text{ m}^2 \text{N}^{-1}$

derived from Blackslee *et al.* (1970), he attempts to deduce S_{44} using the uniform stress and his own 'uniform strain' treatment from results on Thornel fibre, determining the corrected modulus on the assumption that the effect of the pores is restricted to a reduction of the effective cross-section. He finds that the value of S_{44} deduced varies greatly, though uniformly, with the preferred orientation – from 0.4 to 0.9 $\times$ 10^{-10} m^{2}N^{-1} for the second model. He deduces that both models are inappropriate, and proposes a new model termed 'elastic unwrinkling' in which it is assumed that the graphitic ribbons are continuous from crystallite to crystallite and so are kinked. A stress on these ribbons will tend to rotate component crystallites and so increase the preferred orientation of the individual layers. The environment or boundary restraint of the ribbons resists the tilt of the layers and at low angles a component of tilt of the stress will cause an elongation of the layer planes. He proposes a simple model in which

$$1/E_c = l_z S_{11} + m_z S^*$$

where

$$l_z = \frac{\int \cos^2 \phi g(\phi) \, d\phi}{\int \sin \phi g(\phi) \, d\phi}$$

$$m_z = \frac{\int \sin^2 \phi g(\phi) \, d\phi}{\int \sin \phi g(\phi) \, d\phi}$$

Ruland shows that S^* is constant at 0.275×10^{10} m^{2}N^{-1}. Incidentally, for an isotropic glassy carbon $l_z = m_z = \pi/4$ and assuming the value for S^*, E_c is predicted to be 2.9×10^{10} N m^{-2}. This should be compared with the similarly corrected measured values of 3.0×10^{10} N m^{-2}. Thus, S^* is apparently constant throughout the full range of preferred orientations from $R_z = 0.66$ to 0.97. This is an extraordinarily accurate prediction for such a simple approach to the dewrinkling process, but it does seem to indicate that, for polymeric carbons at least, we must abandon the classical concept of the overall deformation being composed of the deformation of discrete units and must adopt the concept of continuity of matter from one crystalline region to the next, so that the deformation of one crystalline region is bound up with the deformation of crystal regions beyond its nearest neighbours. This must be true for ordinary crystalline polymers and graphites as well as polymeric carbons.

S^* is approximately equal to S_{33} and it would be tempting to make this identification, but the shear compliance S_{44} must make some contribution. Cooper and Mayer (1971) have made some measurements after short doses of irradiation and claim that the modulus increases by $\sim 5\%$. The scatter of results was greater than 20%. Jenkins and Jouquet (1968) showed that S_{44} in a perfect graphite crystal for a similar dose increases by a factor of 20 and so the shear compliance must only have a relatively small effect in polymeric carbons. The other constants are not affected by such low doses.

A disadvantage of the simple approach is that as l_z tends to 0, m_z tends to ∞, thus making the equation unusable to calculate elastic constants at right angles to the fibre axis. It is, however, possible to modify the original concept to take this into account. Curtis *et al.* (1968) show that the Young's modulus and the preferred orientation of fibres increase linearly and reversibly with the applied stress. This corroborates Ruland's evidence for a dewrinkling mechanism. Ruland predicts that

$$\frac{\mathrm{d}E_c}{\mathrm{d}\sigma} = \frac{\mathrm{d}\ln E_c}{\mathrm{d}\ln l_z}$$

This increases with the orientation factor from 6 to 80, being 20 when $R_z = 0.95$, which is very close to the experimental values.

The similarity with high polymer fibres stretched to the limit of their extensibility should be noted and so should the high elastic strain to fracture, especially in type II fibre.

For fibres with little orientation, Bacon and Smith (1965) have shown that the force–elongation curves are linear with no hysteresis until test temperatures above 1000 °C are attained.

6.3   The effect of heat-treatment and ambient test temperatures

The effect of heat-treatment temperatures on stiffness depends on the degree of preferred orientation in a polymeric carbon. Andrew and Sato (1964) worked on 'hard filler–hard binder' systems made from phenolic resins and also Tokai Glassy Carbon. Their results are shown in fig. 32. It is clear that as the heat-treatment temperature increases above the calcining temperature of 1000 °C, the modulus decreases.

For a high degree of preferred orientation, as exhibited by PAN carbon fibres, the modulus increases continuously with increasing heat-treatment temperature. According to Watt (1970), untreated PAN fibres have initially a straight line stress–strain curve followed by plastic deformation with increasing tangent modulus to fracture – due to increased chain orientation. The oxidized fibre shows the same initial modulus (showing the same mechanism applies) followed by easy

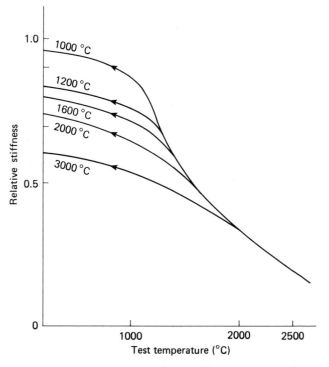

Fig. 32. Variation of relative stiffness with test temperature in glassy carbons heat-treated progressively from 1000 to 3000 °C (after Andrew and Sato, 1964).

plastic flow up to fracture – presumably because no further orientation can take place at room temperature after the completion of the oxidation process. It also shows that the hydrogen-bonds responsible for inter-molecular forces are absent in the oxidized ladder polymer. After pyrolysis to 600 °C the modulus increases due to chain coalescence. In common with all carbonization processes, spatial factors are effective. For a fixed pyrolysis temperature the modulus reaches saturation after about 10^3 s. On raising the temperature by 100 °C another 10^3 s are required to reach another saturation value. When this carbonization process is complete at about 1000 °C the modulus has risen by a factor of 20 from 10 GN m^{-2} to 200 GN m^{-2}, as illustrated in fig. 33 (curve A).

Heat-treatment above 1000 °C increases the modulus progressively to 420 GN m^{-2}. This is clearly different from the effect in isotropic glassy carbon. Using Ruland's dewrinkling model this can be interpreted as follows. As the annealing temperature is increased above 1000 °C, the crystallite size (L_a) increases and so presumably the distance between inter-ribbon cross-links increases. The environmental forces preventing dewrinkling must, therefore, decrease; so S increases. Simultaneously an increase in heat-treatment temperature must 'iron out' small wrinkles between strong cross-links to allow L_a to increase. Thus the effective S_{11} decreases. For randomly oriented ribbon, increases in S will super-sede decreases in S_{11}. For a high degree of orientation, a decrease in the effective S_{11} will overcome any increase in S^*.

For the tensile testing of our surface-oriented phenolic glassy carbon fibre, small uniform lengths were mounted vertically in resin and the diameter of the polished cross-section measured using an optical microscope (Kawamura and Jenkins, 1970). Accurate measurement of diameter was very difficult especially for fibres with diameters less than 10 μm. Both ends of a test sample were fixed with resin on a cardboard frame, both sides of which were cut before measurements were made.

Figure 33 illustrate the variation of Young's modulus of a thick fibre ($\sim$30 μm diameter) (B) as a function of heat-treatment temperature, compared with that of high modulus fibres of PAN (A) and bulky glassy carbon (C). The PAN data is derived from Moreton *et al.* (1967). The average tensile modulus of fibre heat-treated at 300 °C is 5×10^9 N m^{-2} which is a reasonable value for the Young's modulus of an organic polymer. After heat-treatment at 350 °C, the Young's modulus increases by about 20% to reach 7×10^9 N m^{-2}, due to the formation of inter-molecular cross-links between the chain molecules. The Young's modulus is almost constant between 350 and 500 °C but increases rapidly thereafter to reach a maximum value of 63×10^9 N m^{-2} at 1500 °C. The rapid increase in the Young's modulus between 500 and 1500 °C

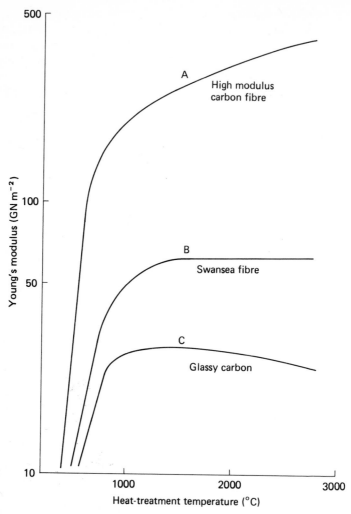

Fig. 33. Variation of the axial Young's modulus of phenolic resin fibres on pyrolysis, compared with polyacrylonitrile fibres (*A*) and phenolic resin blocks (*C*).

corresponds to a rapid increase in the elastic modulus of isotropic bulk material with the same heat-treatment.

At heat-treatment temperatures above 1500 °C the elastic modulus of isotropic material decreases considerably; the modulus of the fibre keeps constant, owing to surface orientation in the fibres. Presumably, the effect of surface orientation is negligible when the diameter of the fibre is large and heat-treatment temperature is low.

When the fibres are heat-treated above 1500 °C the elastic modulus of the isotropic core decreases gradually, as in curve *C*, fig. 33, whereas the modulus of the textured sheath increases, as in curve *A*. Therefore, any decrease in the elastic modulus of the fibre core is compensated for by increases in the thickness of the textured sheath which has a much higher value of Young's modulus along the fibre axis. This is manifested in a constant value of the modulus of the fibre between 1500 and 2500 °C, although the stiffness of glassy carbon blocks shows a considerable decrease over this heat-treatment range.

Figure 34 shows that the Young's modulus increases gradually with decrease of diameter. This is explained by an increase in the degree of molecular orientation with a decrease in diameter. The thickness of the oriented layer is almost independent of the fibre diameter, which means that the ratio of the oriented layer to the fibre diameter increases with decrease in fibre diameter. Since the oriented layer has a much higher elastic modulus than the isotropic core, the Young's modulus of the fibre increases with decrease in fibre diameter.

A similar result has been obtained by Bacon and Smith (1965) with carbonized rayon fibres. They also claimed that the orientation in the original Rayon fibres is memorized in the resulting carbon fibres and the behaviour of the Young's modulus of the fibres heat-treated at high

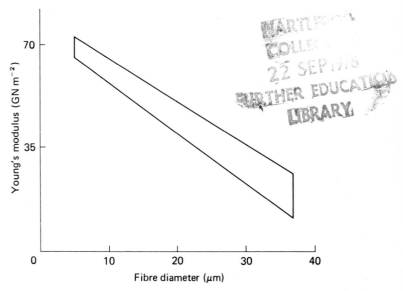

Fig. 34. Variation of the Young's modulus of carbon fibres with fibre diameter.

temperatures is due to an increase in orientation along the fibre axis. Although they could not show direct evidence of the memorized preferred orientation in the carbonized rayon fibre, it is inferred that the induced orientation is memorized in the surface.

Jones and Duncan (1971) have shown that the Young's modulus of carbon fibres derived from PAN and rayon is also dependent on the fibre diameter. This, again, is attributed to the presence of a non-typical layer of surface material.

Young's modulus of the isotropic carbon fibres made from an oil–pitch system increases with decrease of fibre diameter (T. Araki, private communication). This is also explained by preferred orientation in the surface, which is common in carbon fibres and is thought to be induced during melt extrusion and spinning.

Andrew and Sato (1964) showed that the elastic modulus of 'hard filler–hard binder' carbons decreases as the test temperature is raised (cf. fig. 35). Yamada (1968) reports a similar effect in pure glassy carbons. By contrast, soft electrographites always exhibit marked increases in Young's modulus. This is attributed to the presence of microcracks within the highly layered grains formed during cooling from the firing temperature as a result of stresses between anisotropic grains. As the temperature is increased the cracks fill, the differential movement between material above and below the cracks is inhibited and so the material stiffens. The absence of this positive temperature coefficient of Young's modulus in polymer-carbons is indicative of the absence of interlamellar microcracks, presumably because of the very small layer sizes and high degree of cross-linking. High internal stresses must be present, but these do not affect the effective modulus.

The Young's modulus of a porous polymeric carbon, cherry charcoal, has been measured as a function of test temperature for various heat-treatment temperatures (D. R. Blankenhorn, private communication). Typical plots are shown in fig. 35. Clearly, the effect of heat-treatment is first to drastically lower the modulus and decrease its thermal dependence. Above 500 °C, the modulus rises rapidly and thermal dependence remains low. The slope is nevertheless negative.

6.4   Damping and visco-elastic behaviour

Damping in carbons is highly dependent on the width of free sheet (L_a). Thus polymeric carbons have very low damping even when annealed at high temperatures. In contrast, graphitic carbons exhibit very high strain dependent damping after annealing at graphitizing temperatures. Tsuzuku and Kobayashi (1961) reported that the internal friction of Japanese glassy carbon is low and about 10% of that of commercial

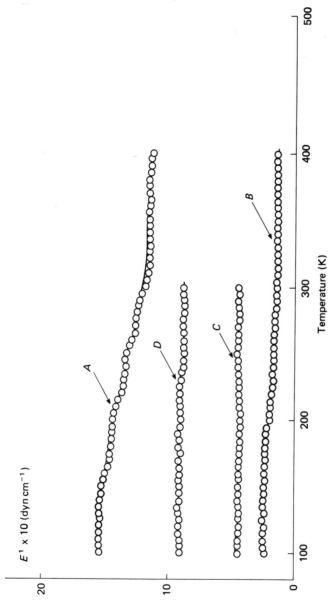

Fig. 35. Variation of the dynamic elastic modulus (E) of a cherry wood charcoal with test temperature at various stages of its pyrolysis. A, original wood; B, 570 K; C, 870 K; D, 1170 K.

polycrystalline graphites. They found that the internal friction of glassy carbon is almost independent of strain amplitude, examined between 0.0001 and 0.1 %. They also studied the temperature dependence of the internal friction of glassy carbon between liquid nitrogen temperature and room temperature, and observed two peaks at −90 and −40 °C, but no accompanying modulus decrement.

Tsuzuku (1964) studied the internal friction of Japanese glassy carbon which was subjected to a neutron radiation dose of about 10^{15} nvt. Although no change in dynamic modulus was observed, within the limits of experimental error, he found that the internal friction increased about 10 % and that the strain-amplitude dependence also increased.

Taylor and Kline (1967) measured the internal friction of French glassy carbon V10 (heat-treated at 1000 °C) at ambient temperatures between 4 and 570 K. They observed low internal friction values, but the two peaks found by Tsuzuku and Kobayashi were not reproduced. They reported that the dynamic modulus of V10 decreased with increase of ambient temperature. The internal friction of V10 after heat-treatment at 1800 °C showed a relatively large peak at 430 K, at which temperature the dynamic modulus decreased rapidly. They also reported that the dynamic modulus increased by about 0.2 % at room temperature after the sample was irradiated to a neutron dose of 1.4×10^{17} nvt. The neutron irradiated glassy carbon showed an internal friction peak at 540 K. However, their experimental data fluctuated widely over several experimental runs for each sample. It is difficult to determine whether this fluctuation is due to the experimental errors or to the true effect of the experimental run on the internal structure of glassy carbon.

The effect of various heat-treatment temperatures on the damping of pyrolysed cherry wood has been measured using thin cylindrical specimens pyrolysed by us (D. R. Blankenhorn, private communication). This is illustrated in fig. 36. In the plot of damping against test temperature, the peak at ~240 K is the most prominent feature. Heat-treatment at 590 K causes this to increase. Further pyrolysis to 870 and 1170 K causes the damping to diminish and the peak to migrate to progressively lower temperatures. At the highest temperature of pyrolysis the peak has drifted to 130 K and the room temperature damping (Q^{-1}) has dropped from 16×10^{-3} in the original wood to 7×10^{-3} in the final charcoal.

6.5 High temperature creep

Fischbach (1967) has shown that at high temperatures, above 2000 °C, it is possible to have permanent deformation in glassy carbon which

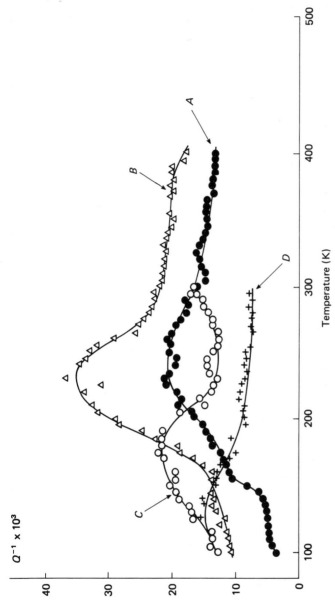

Fig. 36. Variation of the internal friction (ϕ) of a cherry wood charcoal with test temperature at various stages of its pyrolysis. *A*, original wood; *B*, 570 K; *C*, 870 K; *D*, 1170 K.

causes anisotropy as determined by magnetic susceptibility measurements. He also shows that 7% elastic recovery is possible. It would seem that above this temperature the material becomes visco-elastic and 2500 °C can be regarded as a 'glass point', by analogy to the behaviour of low temperature polymers. The plasticity is accompanied by an increase in open porosity as is observed in pyrolytic carbon and pitch–coke graphite.

Unoriented carbon fibres (cf. Bacon and Smith, 1965) show permanent plastic deformation at 1900 °C. As the temperature is raised from room temperature, the strain–stress curves begin to show some hysteresis at 1200 °C, this becoming greater as the temperature rises up to 1900 °C. The room temperature modulus invariably increased as a result of these high temperature tests and it was inferred that permanent reorientation of the ribbons had taken place.

Such permanent aligning of ribbons was later confirmed by Bacon and Schalamon (1967) who showed that carbon fibres from pyrolysed rayon could be pulled by 300% at 2750 °C. As the strain is increased the room temperature modulus increases from 70 GN m^{-2} to 630 GN m^{-2}. Johnson (1970) shows that a similar effect is observed in PAN carbon fibre, for which 27% strain produces a fibre modulus of 670 GN m^{-2}.

Hawthorne *et al.* (1970) show that carbon fibres made from refined asphalt can be pulled at 2800 °C to extensions of 140%. The effect of the high temperature axial plastic deformation is again to preferentially orient the component crystallites, aligning the ribbon planes parallel to the fibre axis. Young's modulus increases from 30 GN m^{-2} to 440 GN m^{-2}. It is concluded that plastic deformation involves the removal or transfer of cross-links between ribbons allowing dewrinkling or dekinking, producing a preferred orientation in the direction of extension.

Fischbach (1969) has measured the tensile creep of GC-20 and GC-30 glassy carbon using the usual dog-bone shape. The reduction in cross-sectional area of GC-30 was nearly that predicted for constant volume deformation; the reduction in area of GC-20 was a little smaller. By changing the stress during test, the dependence of the isothermal creep rate on elongation was determined at various stress levels. The creep rate decreased monotonically with elongation under constant load and temperature. The stress dependence of the creep rate above 2500 °C could be represented by:

$$\dot{\varepsilon} = A\sigma^n$$

where $n = 1 + B\sigma$ and $B = 5 \times 10^{-8}$ m^{2}N^{-1}.

Tensile elongation produced preferred orientation. Creep recovery occurred when the stress was reduced. Although only a small recovery resulted from small stress reductions, at least 25 % of the creep elongation was recoverable when the stress was reduced to zero. Of the recorded elongation of over 10 %, 2 % was recovered instantaneously and a further 2 % after 30 min. By changing the temperature at constant load, the dependence of creep-rate on elongation was determined at various temperatures. The effective activation energy determined from these results was 350 kcal mole^{-1} above 2500 °C. This is much higher than the activation energy of ~250 kcal mole^{-1} which is now 'well established' for pyrolytic carbon and coke–pitch composites, and is related to the energies required to form and move lattice vacancies in the layer planes of graphite. It is possible that the high observed activation energy for glassy carbon is only an apparent or effective value and cannot be related to the physical processes involved.

6.6   Strength and fracture

The elastic constants are intrinsic properties of the material. Strength, however, is also sensitive to defects. The effect of such defects is much more pronounced in glass-like carbon than in polycrystalline graphite because of the inability of the former to relax the stress build-up at inhomogeneities.

Fischbach and Kotlensky (1965) measured the tensile and structural properties of GC-20 and GC-30, showing that the tensile strength at room temperature was 42 MN m^{-2} for GC-20 (heat-treatment temperature 2000 °C) and 40 MN m^{-2} for GC-30 (heat-treatment temperature 3000 °C). As the test temperature was increased, the strength increased; at 2500 °C it was over 140 MN m^{-2}. Glassy carbon thus shows the same strength behaviour as coke–pitch aggregates in that the strength increases with test temperature.

Sato and Asakura (private communication to Yamada, 1968) have determined the strength of bulk Tokai glassy carbon as a function of heat-treatment temperature and the results are presented in table 10. The inference is made that heat-treatment above 1000 °C drastically lowers the tensile strength of glassy carbon. The discrepancies between room temperature strengths measured by the two laboratories may be ascribed to differences in specimen preparation and testing technique.

By measuring the strength variation with gauge length, Moreton (1969) shows that the strength of PAN carbon fibre is sensitive to defects. For 100 mm gauge length, the mean strength is 1.86 GN m^{-2}, increasing to 2.1 GN m^{-2} for 50 mm gauge length and 2.8 GN m^{-2} for 5 mm gauge

TABLE 10   *Compressive and flexible strengths of glassy carbon (round-bar specimens).*

Test	Grade	Average value (MNm^{-2})	Standard deviation (MNm^{-2})	Heat-treatment temperature (°C)
Compression	GC-10	369	130	1000
	GC-20	200	43	2000
	GC-30	95	15	3000
Flexion	GC-10	165	64	1000
	GC-20	132	27	2000
	GC-30	121	30	3000

length. Extrapolation to 1 mm indicates a possible strength of 3.5 GNm^{-2}. It has been shown that some of these defects are voids or inclusions in the parent fibre carried over to the carbon fibre (Johnson, 1969). The identification and counting of various flaws visible in the parent fibre under the optical microscope leads to the conclusion that improvement in fibre cleanliness will increase the carbon fibre strength.

Johnson (1969) has shown that etching the PAN carbon fibres produced appreciable strength increases, especially for carbons annealed at less than 1200 °C. Above about 900 °C for the etched fibres and 1250 °C for the unetched fibres, the strength began to decrease, data points for both falling on the same curve. For fibres treated at temperatures below 1000 °C, the failure was essentially brittle and was governed by the presence of stress-raising defects, many of which were actually identified on fracture surfaces. The effect of etching was to remove some of these defects and so improve the strength of the fibre.

It is generally agreed that the strengthening and stiffening of polymeric carbons up to 1000 °C is due to the increase of cross-links between ribbon units. Above 1200 °C the fibre strength is independent of whether the fibres are etched or not, indicating that intrinsic defects rather than surface flaws govern the strength thereafter. Cooper and Mayer (1971) suggest that these faults may be a result of the increase of ease of shearing inside the crystallites with growing L_a. They then invoke the Hall–Petch relationship for the effect of grain size on the yield stress of poly-crystalline metals. Naively, they refer to dislocation pile-ups in glassy carbon – even though the maximum crystal size is only twice the calculated width of the core of one partial screw dislocation. However, the basic suggestion is sensible in that the separation between two perfect sheets can be regarded as a thin crack, the stress on the boundary increasing with L_a for a given applied stress.

Jones and Johnson (1971) have attempted to measure the intrinsic strength of high modulus carbon fibres, i.e. that free from defects, by pulling looped single fibres derived from PAN. They conclude that the intrinsic strength is probably in excess of 7 GN m^{-2} after heat-treatment at 1600 °C.

Williams *et al.* (1970) used the same technique with rayon fibres as precursors to show that the strength increases linearly with the Young's modulus, the slope being 0.005. They are led to reject the Griffith brittle fracture criterion for their material, in favour of that of Marsh (1964) who assumed an elasto-plastic model for silicaceous glass. Failure is predicted at a critical 'yield' stress in a region of density fluctuation in a material which shows no work-hardening.

It is expected that fibres with smaller diameter have higher ultimate tensile strength, because the concentration of defects should be lower. Figure 37 illustrates the variation of the ultimate tensile strength of glassy carbon fibre, derived from phenolic resin and heat-treated at 900 °C, as a function of diameter. The strength increases rapidly with decrease of diameter, especially for fibres with diameters smaller than 15 μm. This relationship between the ultimate tensile strength and the fibre diameter is very similar to that of glass fibre, which suggests that the fracture mechanism is the same in both and that the main factor affecting

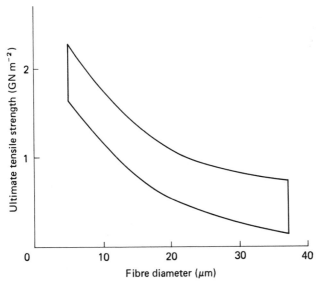

Fig. 37. Variation of the ultimate tensile strength of glassy carbon fibre with fibre diameter.

the strength of the fibres is the presence of defects such as voids and cracks.

It is reasonable to assume that thicker carbon fibre has a higher concentration of internal flaws, because thinner fibre has a larger ratio of free surface area to volume and, therefore, escape of gases during carbonization is easier. In thick fibres, internal gas pressure is increased during carbonization and this results in the creation of cracks and voids. Indeed, large bulk samples of glassy carbon cannot be made without producing macropores and microcracks.

After spinning, the molecules in the fibres tend to coil up by heat motion and this results in shrinkage, which is opposed by tension supplied from the bobbin in spun fibres which exhibit much reduced plasticity during cooling. This could be responsible for the formation of surface microcracks. Molecular retraction is less significant in thinner fibre because it is cooled faster, which leads to smaller axial shrinkage. The strength of fibre immediately after extrusion is quite high, but decreases considerably with time during cooling. This supports the proposition that surface cracks are formed during cooling.

It is difficult to discuss the formation of mechanically induced surface damage because this is formed less systematically. Analysis is possible only for flaws due to the extrusion process which produces surface cracks aligned parallel to the fibre axis, and thus renders them less effective.

The ultimate tensile strength of the fibre is proportional to the reciprocal of the diameter, as expressed by the following equation:

$$\sigma = \sigma_0 + K/D$$

where D is the fibre diameter and σ_0, K are constants. In the present case,

$$\sigma_0 = 2 \times 10^8 \, \mathrm{N\,m^{-2}}, \qquad K = 11 \, \mathrm{N\,m^{-1}}$$

where D is measured in metres. This relationship is very similar to that for glass fibres reported by Griffith (1920).

σ_0 represents the ultimate tensile strength of bulk glassy carbon. The maximum ultimate strength of bulk glassy carbon has been found by Fitzer and Schäfer (1970) to be $1.5 \times 10^8 \, \mathrm{N\,m^{-2}}$. This is somewhat smaller than that predicted by extrapolation, probably because of difficulties in machining bulk glassy carbon without producing serious surface damage.

By plotting the reciprocal of the ultimate tensile strength against the fibre diameter, it is possible to estimate the limiting tensile strength of the fibre by extrapolating the linear relationship to zero diameter.

This gives a tensile strength of 3×10^9 N m^{-2} which corresponds to the theoretical tensile strength of bulk glassy carbon, assuming the theoretical strength to be $\frac{1}{10}$ Young's modulus ($\sim 30 \times 10^9$ N m^{-2}).

Watt and Johnson (1970) have shown that the strength of carbonized PAN fibres increases with Young's modulus only up to 1500 °C. Further heat-treatment to higher temperatures causes a marked deterioration in strength. In the case of carbonized rayon fibres (Bacon and Schalamon, 1969), the strength continues to rise in step with the modulus up to the highest heat-treatment temperatures.

By way of explanation for the anomalous behaviour of carbon fibres, Jones and Duncan (1971) postulate the presence of a 'core' material which is distinguishable from the 'sheath' of a fibre. If the thermal expansions of these two components differ, heat-treatment to high temperatures (and cooling therefrom) will promote the opening up of microfissures which lower the strength. Sharp and Burnay (1971), on the other hand, have observed 'diconic' cavities derived from the volatilization of inorganic impurities after heat-treatment above 1800 °C. They claim that these are the cause of the observed weakening of carbon fibres derived from PAN.

The variation of the ultimate tensile strength of thick glassy carbon fibres with heat-treatment temperature is illustrated in fig. 38 (Kawamura and Jenkins, 1972). The fact that the ultimate tensile strength increases continuously up to a heat-treatment temperature of 1000 °C may be attributed to an increase in Young's modulus. However, the increase in Young's modulus is about three times as large as that in the tensile strength, presumably because the concentration of defects also increases during carbonization. After heat-treatment above 1000 °C, the tensile strength of the fibre decreases with increase in temperature, although the Young's modulus stays constant. Presumably, when some of the inter-ribbon cross-links are broken by high temperature heat-treatment, the apparent crystallite diameter increases gradually, accompanied by release of internal strain energy in the graphitic molecules. Increase in crystallite diameter makes crack propagation easier. The severance of interlayer cross-links, in particular, promotes such crack propagation, leading to a fall in tensile strength at high temperatures.

Both ultimate tensile strength and Young's modulus increase with increase of heat-treatment temperature between 500 and 1000 °C, but the elongation at fracture of fibres shows a minimum at a heat-treatment temperature of 700 °C, as indicated in fig. 39. Thus brittleness increases drastically for heat-treatment temperatures between 500 and 700 °C, and fibre heat-treated at 700 °C is very difficult to handle. An increase in brittleness may be connected with a closer equality of strength

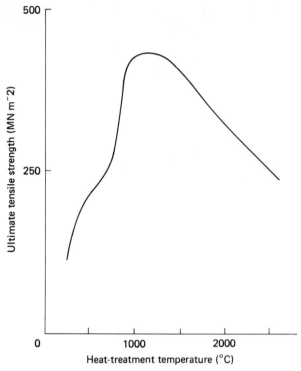

Fig. 38. Variation of the ultimate tensile strength of glassy carbon fibre with heat-treatment temperature.

of bonds in various directions with the formation of cross-links of covalent character between the aromatic layers in the direction perpendicular to the plane of the ribbon molecules. At higher temperatures, there is an increase in the anisotropy of bond strengths perpendicular and parallel to the plane of the aromatic ribbon molecules. This leads to the increase in the strain to fracture observed at heat-treatment temperatures above 700 °C.

The strength of pyrolysed wood is greater than is commonly believed. For instance, consider the pyrolysis of lignum vitae which has a compressive strength of 70 MN m^{-2}. Measurements made in our laboratory show that this strength drops drastically to a value of 10 MN m^{-2} after pyrolysis at 400 °C. Further pyrolysis to 1000 °C, however, causes the compressive strength to rise to 60 MN m^{-2} – almost the same as the original wood value, despite a large weight loss and a reduction in density to 0.9 g cm^{-3}. Such material has been used for medical implant work to measure tissue ingrowth into a porous carbon (Professor A. O.

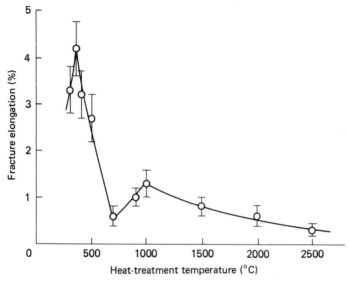

Fig. 39. Variation of strain to fracture of glassy carbon fibre with heat-treatment temperature.

Mack, Eastman Dental Hospital, London). Interestingly enough, if this porous material is impregnated with a resin, the strength leaps to 190 MNm⁻².

6.7   Hardness of glassy carbon

Glassy carbon is a very hard material; it will scratch most forms of silicaceous glass. Hardness and resistance to deformation are sensitive to the amount of cross-linking between chain polymers and between the aromatic ribbon molecules. It is therefore possible to elucidate the carbonization mechanism and the structure of the resulting glassy carbon from hardness measurements of the resin carbonized at various temperatures.

A Vicker's hardness testing machine can be employed for this purpose. The Diamond Pyramid Hardness (DPH) number is determined from the average length of the two diagonals in the indentation of a four-sided diamond pyramid under an applied load of typically 10 kg.

We have found (Kawamura, 1971) that the DPH is a useful measure of the Young's modulus of our carbons. This is illustrated in fig. 40 for materials subjected to a range of heat-treatment temperatures. It may be noted that the relationship between Young's modulus and DPH is linear, intersecting both axes at zero.

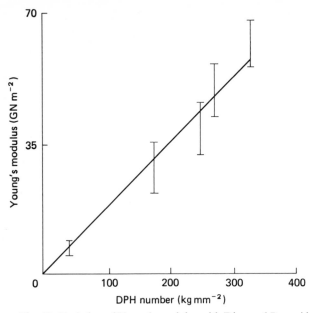

Fig. 40. Variation of Young's modulus with Diamond Pyramid Hardness for a range of glassy carbons subjected to various heat-treatment temperatures.

Figure 41 illustrates the variation of hardness as a function of heat-treatment temperature. Some of the hardness values of commercial glassy carbons determined under the same conditions are also shown for comparison. Swansea glassy carbon shows lower hardness values than commercial glassy carbons and the difference becomes considerable for samples heat-treated at very high temperatures.

The hardness increases by about 20% after heat-treatment at 350 °C and decreases slightly up to 500 °C. Above 500 °C, it rises rapidly to reach a maximum at 1500 °C and then decreases considerably. In particular, a marked decrease is observed at 2700 °C. The hardness of the sample heat-treated at 2700 °C is much lower than that of GC-30 heat-treated at 3000 °C, even though X-ray diffraction studies reveal that the carbons have almost the same lattice parameters and crystallite size. The variation in hardness between the two types of glassy carbon is considered to be due to the nature and concentration of cross-links present in these carbons (Kawamura, 1971).

The increase in hardness after heat-treatment at 350 °C supports the conclusion, deduced from infra-red spectroscopy studies, that inter-molecular cross-links are then formed between the chain polymers.

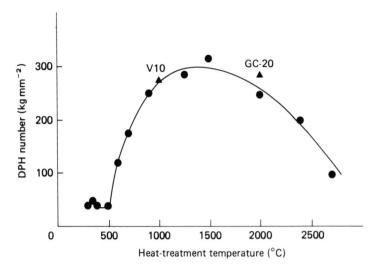

Fig. 41. Variation of the hardness of glassy carbon as a function of the heat-treatment temperature.

The decrease in hardness observed between 350 and 500 °C indicates that intermolecular forces decrease at these temperatures. Infra-red spectroscopy studies suggest that aliphatic ether linkages are then ruptured. The decrease in hardness must be associated with this scission of aliphatic ether linkages, resulting in a more open structure.

The rapid increase of hardness on annealing above 500 °C is due to the formation of inter-ribbon cross-linking as implied in previous sections. Figure 42 shows the variation of hardness as a function of hydrogen/carbon atomic ratio. The hardness increases approximately linearly with decrease in this ratio, which means that the intermolecular forces increase with an increase in the number of inter-ribbon cross-links after dehydrogenation.

The decrease of hardness for heat-treatment temperatures above 1500 °C indicates that rearrangement of the crystallite in the initial stage of graphitization takes place at these temperatures. As we have indicated previously, inter-ribbon cross-links are broken between 1500 and 2000 °C, allowing the release of internal strain energy in the tangled ribbons, leading to a decrease in the intermolecular forces and the associated hardness.

It is interesting to note that material heat-treated at 400 °C shows marked delayed elasticity or elastic after-effect. The diagonals in the indented shape disappear after one week and only a small hole in the

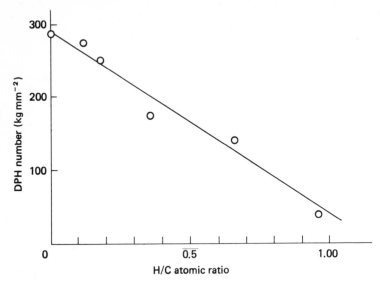

Fig. 42. Variation of the hardness of glassy carbon with the hydrogen/carbon atomic ratio.

centre remains. A sample heat-treated at 500 °C showed similar recovery, but its recovery rate was much smaller. The occurrence of high elasticity (the capacity for sustaining large elastic strains) suggests a loose, open polymeric network structure which is consistent with our explanation for the decrease in hardness at these heat-treatment temperatures.

In samples heat-treated above 700 °C, the indentation depth is almost completely recovered immediately after removal of the applied load as long as the material is not fractured. No obvious evidence of delayed elasticity can be observed at room temperature. The material is, therefore, converted into a three-dimensionally cross-linked close network structure at these temperatures.

7 Chemical reactivity

7.0 Generalities

There are three ways in which polymeric carbons may be attacked. Oxidation takes place in air at the carbon surface and carbon is removed as its volatile oxides. Since there is no accessible internal surface in glassy carbon it is relatively inert. Many polymeric carbons can be 'activated' by such oxidation at various stages of the carbonization process. The result is an open structure with an extremely high internal surface which is extremely reactive.

Sodium and potassium attack polymeric carbon by penetration between the graphitic layers – a process termed intercalation. In isotropic material the result is a violent disruption.

Other metals attack carbon surfaces readily at higher temperatures, dissolving the carbon atoms and reprecipitating them as graphite crystals.

The chemical behaviour is a clear manifestation of the presence of stacked ribbons – witness the ease with which potassium and sodium are intercalated. The rapid high temperature attack and easy wetting with other metals shows the presence of easy penetration paths along ribbons. At low temperatures the difficulty of reaching internal surfaces hinders attack by oxygen and corrosive chemicals.

7.1 Oxidation

The rates of oxidation of an impermeable glassy carbon in oxygen, carbon dioxide or water vapour have been found by Lewis (1965) to be lower than those of any other carbon. This is illustrated in fig. 43 which shows 'Arrhenius' plots of the logarithm of reaction rate against the reciprocal of absolute temperature, for various carbons in dry air after pre-heating to 900 °C in argon to remove adsorbed gases. It should be noted that the activation energies of all the carbons are very similar ($\sim$13 kcal mole^{-1}) even though the electrographite is a highly graphitized material.

Work on the kinetics of burn-off (Carmen Navarro, private communication) has shown that the activation energy is constant for pyrolysed phenolic resin heat-treated at temperatures between 500 and 1000 °C (see Arévalo and Jenkins, 1974). Only the pre-exponential rate factor

135

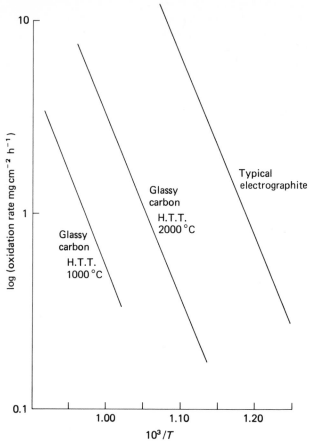

Fig. 43. Arrhenius plots of the rate of oxidation with temperature of 'vitreous carbons' VC1 (curve 1) and VC2 (curve 2), compared with a typical electrographite (curve 3) (after Lewis, 1965).

in the Arrhenius relationship is affected – by several orders of magnitude. This indicates that basic, rate-controlling chemical reaction is independent of heat-treatment temperature; only the permeability to the attacking gas is affected. It is also reported that small additions of phosphorus and boron to the original resin lower the rate constant further.

Glass-like carbon is remarkably inert to low temperature oxidation via a liquid medium. Normal graphite plate is destroyed to a powder at room temperature by a mixture of concentrated sulphur and nitric acids in 40 h. Glassy carbon shows no change in weight or volume after 5 months in such conditions. Even at 120 °C, the weight loss was only 0.5% after 6 h (Lewis, 1965). Vitreous carbon crucibles are eventually attacked by repeated sodium peroxide and potassium bisulphate fusions, but, of course, are not susceptible to the gross metallic contamination of the melt inherent in using nickel or iron crucibles. Porous polymeric carbons behave very differently because of the enormous increase in available porosity.

7.2   Activated carbons and accessible porosity

'Active carbon' represents a very important usage for polymeric carbons. For many centuries they have been used to decolourize liquids and to purify water. Today, they have many and varied uses: water purification, adsorption of gases, recovery of volatile liquids, pharmacology, chromatography.

There is a wide variety of starting materials, chosen mainly for availability and cheapness: hard or soft woods, nutshells – especially those of coconuts – sugar, as well as man-made polymers. The carbon is not graphitizable in every case and the nature of the precursor determines to a large extent the final pore texture. This porous texture is not created by subsequent gaseous activation, e.g. by steam at 700 °C, but is only thereby exposed. Graphitizable carbons have only a very fine inner pore texture which can not be opened up by subsequent activation.

Carbonized polyvinylidene chloride (Saran) is a good example, possessing exceptional adsorption capacity. Dacey and Thomas (1954) showed that for pyrolyses at 700 °C the volume adsorbed at saturation was close to $0.43 \text{ cm}^3\text{g}^{-1}$ for small molecules such as nitrogen, $0.35 \text{ cm}^3\text{g}^{-1}$ for cyclohexane and $0.27 \text{ cm}^3\text{g}^{-1}$ for tetraethylmethane. The distribution of pore dimensions is therefore remarkably uniform and the product is a good material for studying the evolution of pore texture in porous substances carbonized at high temperatures. Even without any activation, Saran carbon prepared at 700 °C contains a nominal surface area with respect to nitrogen of $1270 \text{ m}^2\text{g}^{-1}$; activation increases this by a factor of 4.

The production of porosity during the pyrolysis of pure polyvinylidene chloride was studied by Dacey and Cadenhead (1960). The variation of the discrepancy between the density in helium and the density in mercury was determined and compared as a function of the weight loss. For a weight loss up to 30 % there was very little discrepancy but for greater weight loss the discrepancy rose rapidly to an equivalent porosity of $0.4 \text{ cm}^3\text{g}^{-1}$ at 75 % weight loss. Marsh and Wynne-Jones (1964) have determined the variation of specific surface area in nitrogen and carbon dioxide as pyrolysis proceeds in PVC and PVDC. The porosity of pyrolysed PVC was found to be extremely fine; the material is practically impervious to nitrogen at low temperatures and is impermeable to all gases after heat-treatment to 800 °C.

The specific surface area curves for PVDC pyrolysis and those of other polymers forming non-graphitizable chars, such as polydivinyl benzene, polyfurfuryl alcohol, cellulose and melamine–formaldehyde resin, all differ as indicated in table 11. The specific surface area in PVDC carbons measured with both nitrogen and carbon dioxide reaches a maximum of 1400 $m^2 g^{-1}$ at 1600 °C and thereafter decreases.

Polymers producing graphitizable carbons always have a small specific surface area and both the mercury and helium densities approach 2.2 at 2700 °C which means that no closed pores remain. The polymers producing non-graphitizable carbons have high surface areas at 700 °C heat-treatments. At high temperatures the helium density is only 1.4–1.7 $g cm^{-3}$ and the mercury density only 1.0–1.5. Consequently all these carbons contain a large proportion of closed pores, of the order of 0.2 $cm^2 g^{-1}$ at 2700 °C.

The changes in internal surface area, water adsorption and micropore volume as a function of heat-treatment temperature in a glass-like carbon derived from phenolic resin are shown in fig. 44 (after Yamada, 1968). It is clear that a maximum occurs in all cases at 700 °C and it is at this temperature that the pyropolymer is most susceptible to activation carbon. Above this temperature the ribbon clusters tighten and the intrinsic fine porosity of polymeric carbons is then sealed off.

Kipling *et al.* (1964) noted that the pore volume of the 700 °C pyropolymer of PVC is essentially in the form of bubbles of 2000 Å corresponding to the accumulation of gas in the plastic mass, while the pore dimensions of the non-graphitizable carbons are of the order of 10–20 Å, dictated by the rigidity of the solid and the size of the molecules evolved. The pore spectrum is much more attenuated in cokes than in chars at 700 °C.

The essential requirement of an active carbon is the existence of micropores of about 30 Å diameter which is comparable in dimension to the smallest of molecules. The total volume of these micropores may be quite small, 0.10–0.50 $cm^3 g^{-1}$, but their internal surface can be very large (of the order of 100 $m^2 g^{-1}$) and this constitutes the essential quality of an active carbon. This micropore surface is difficult to determine because classical methods tend to overestimate due to capillary effects which are always present at relative pressures of the order of 0.05 to 0.10.

It is often said, implicitly or explicitly, that active carbons contain cylindrical pores, thus simplifying the determination of pore dimensions. Thus for a pore specimen with a pore volume of 0.60 $cm^3 g^{-1}$ and an

TABLE 11   *The development of porosity during pyrolysis.* (After Kipling *et al.*, 1964.)

Precursor		Heat-treatment temperature (°C)	Specific surface area: N_2 at -196 °C ($m^2 g^{-1}$)	Helium density (gcm^{-3})	Mercury density (gcm^{-3})
Polymers producing graphitic carbons	Polyacenaphthalene	700	0.52	1.68	1.58
		2700	0.34	2.13	1.94
	Polyvinyl chloride	700	0.58	1.85	1.68
		2700	0.71	2.21	1.81
	Polyvinyl alcohol	700	1.60	1.77	1.67
		2700	0.42	2.17	1.97
Polymers producing polymeric carbons	Styrene–divinyl benzene co-polymer	700	0.02	1.78	1.40
		2700	0.10	1.49	1.47
	Polyfurfuryl alcohol	700	0.03	1.86	1.44
		2700	0.13	1.45	1.43
	Polyacrylonitrile	700	0.34	1.90	1.51
		2700	0.40	1.57	1.44
	Phenol–formaldehyde	700	26	1.93	1.48
		2700	0.06	1.52	1.52
	Cellulose	700	408	1.90	1.10
		2700	2.23	1.56	1.13
	Polyvinyl fluoride	700	643	1.98	1.02
		2700	0.40	1.20	1.21
Polymers initially pyrolysed at 250 °C in air	Polyvinyl chloride	700	0.16	1.95	1.25
		2700	0.17	1.49	1.34
	Polyacrylonitrile	700	100	1.87	1.45
		2700	0.73	1.57	1.43
	Styrene–divinyl benzene co-polymer	700	523	1.81	1.31
		2700	0.22	1.41	1.34

internal surface area of 1200 $m^2 g^{-1}$ it is calculated that the total pore structure is equivalent to a cylinder 2 nm in diameter and 2×10^8 km long per g of specimen. The pores are considered to be arranged in a network resembling a honeycomb. The thickness of the walls separating two neighbouring cylinders is reckoned to be only 4 Å, a result which is difficult to reconcile with X-ray evidence.

The alternative texture is of crossed cylinders of material, the pores being the interstices necessarily present between these rods. If one presumes a lamellar microtexture a simple calculation indicates a mean distance of 10 Å between the parallel walls of the capillaries which are in reality smooth-walled slots. This is close to the picture derived from X-ray evidence. On heating active carbons, one observes a diminu-

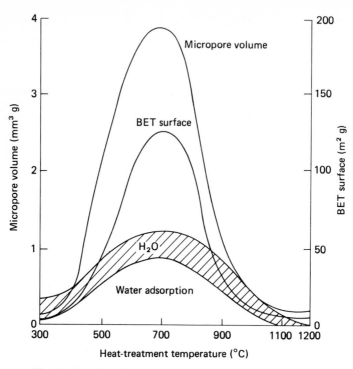

Fig. 44. Variation of micropore volume, BET surface and water absorption of a phenolic resin pyrolysed between 300 and 1200 °C (after Yamada, 1968).

tion in specific surface area simultaneously with crystal growth which is only explicable in terms of a lamellar microtexture.

The hypothesis that micropores consist of slit cracks with approximately parallel walls has been defended in particular by Wolff (1958). He considers active carbons as formed of stacks of parallel carbonized layers separated by more than 3.36 Å and with diameter close to 40 Å, bounded with hydrogen or other functional groups. Macropores will be the interstices between these particles, micropores being closed fissures smaller than 10 Å wide between carbonized layers within the particles. Flat or linear molecules thus penetrate active carbons very easily.

This model corresponds well with that which we visualize for polymeric carbons with but a slightly different emphasis. The network structure of microfibrils is rigidly fixed in space by 500 °C, but does not close up at 700 °C when the greatest internal surface area is available. A large proportion of ribbon edge sites are still filled with hydrogen. A high population of grown-in defects are present and so the number of active sites is enormous. The bonds at ribbon edges or at grown-in defects must be subject to severe distortion. Sites are therefore plentiful enough to attract a passing free molecule which, instead of rebounding, is held for a finite time before its kinetic energy allows it to leave. Polymeric carbons suitably activated are therefore effective adsorbents in which the whole of the internal surface is available to small adsorbed molecules. Indeed, in many cases the heat of adsorption can be greater than the heat of liquefaction.

7.3   Alkali metal attack

In well-graphitized materials, alkali metal atoms enter between the layers of graphite which remain unchanged (cf. Ubbelohde and Lewis, 1960). There are recognizable stages characterized by the ratio of the number of metal layers to graphite layers. In stage I, compounds with potassium, for instance, the chemical formula is C_8K, every carbon layer being separated by a potassium layer. In stage II the chemical formula is $C_{24}K$, in which case a potassium layer separates a pair of carbon layers. These stages are detected by the appearance of new characteristic basal plane X-ray reflections which replace the (002) reflections from the original graphite lattice.

Little work has been carried out to determine whether such intercalation with potassium takes place in 'disordered' carbon. Kroger and Ruland (1955) found that only adsorption takes place when wood charcoal is exposed to potassium vapour. X-ray studies also indicated that no lamellar compounds were produced in polyvinylidene chloride char (a typical non-graphitizing carbon). Platzer-Rideau (1960) and Metrot and Hérold (1967) also claimed that intercalation with potassium does not take place in non-graphitizing carbons.

The effect of sodium on non-graphitic carbons is also of interest because it has rather poorer intercalating properties, even with single crystals, forming only an unstable compound having an ideal formula $C_{64}Na$. It has a lamellar structure in which one sodium layer is intercalated in every eight layers of carbon.

In Swansea, it was found at an early stage that all glassy carbon artifacts shattered explosively in the presence of either potassium or sodium vapour (Halpin and Jenkins, 1968, 1969). For instance, it was found that glassy carbon previously heat-treated at 1300 K consistently exploded at (507 ± 2) K. Glassy carbon previously heat-treated at 2800 K disintegrated at (523 ± 20) K. In the first case, the disruption was particularly violent and was only made safe by the use of small specimens.

Particles of glassy carbon debris were jagged with smooth curved surfaces typical of conchoidal fracture. The average size of a fragment was of the order of a millimetre, demonstrating the effectiveness of the reaction. Particles of these dimensions or less did not disintegrate further. Larger pieces were shot away from the reaction zone at the onset of disruption, and these have been useful in characterizing the initial stages of the process.

The disruption in sodium vapour was found to be less violent and required a much higher temperature than that for potassium, probably because of the smaller expansion to be expected with sodium. The final fragments on average are significantly larger than after potassium attack. Close examination of these particles before removal from the apparatus revealed no apparent colour change.

In all cases, as the temperature was decreased the holding time required increased and the violence of the breakup was reduced. It is clear that there is no true threshold temperature below which breakup does not occur – only a time lag which increases as the temperature is lowered.

X-ray studies were carried out to determine the nature of the final products of the reaction with potassium. Intercalation was proved by the disappearance of the (002) peak and the appearance of other peaks. Stage I compounds were identified by the appearance of broad lines at 0.527 and 0.27 nm while stage II compounds gave lines at 0.297 and 0.87 nm. These values may be compared with those obtained from single crystals which give lines at 0.535 and 0.27 nm for stage I and 0.29, 0.44 and 0.87 nm for stage II compounds (Rüdorff and Schulze, 1954). The characteristic copper colour of single crystals of stage I compound (C_8K) was not obtained, but a dark mauve colour was apparent.

Clearly, intercalation is possible in hard non-graphitizing carbons. There is support for this thesis from Hérinckx, Perret and Ruland (1972) who report that intercalation by potassium does take place in highly distorted ungraphitized carbons such as cellulose yarn carbon and Thornel fibre. The fact that full intercalation is possible is a very good foundation for believing that it can be associated with the process of disruption. The evidence for a stage II compound is particularly interesting as it suggests that three-dimensional ordering of the graphite is not a necessary pre-requisite for obtaining the different stage compounds. Perfect crystalline compounds are not formed since no (*hkl*) of the compounds (C_8K, $C_{24}K$, etc.) were observed and the peaks are very broad. We must assume that only imperfect stage I, stage II compounds are produced, and that there will be little periodicity in the arrangement of potassium atoms between the graphite layers.

We have shown that the layers in glassy carbons are pushed apart with ease and apparently return to their original spacing when the potassium is removed. This would indicate the absence of strong inter-lamellar bonding. Primary bonds at boundaries, on the other hand, need not be broken by intercalation and some accommodation may be possible by rotation about such bonds.

If the rate of build-up of intercalated potassium is assumed to vary as the inverse of the time to fracture (t_f) an Arrhenius plot gives a straight line. The rate of accumulation ($R = 1/t_f$) of potassium, presumably intercalated in the glassy carbon, is given by:

$$R = R_0 \exp\ (-\Delta H/kT)$$

where $R_0 = 10^{14}\ \text{s}^{-1}$ and $\Delta H = (1.4 \pm 0.2) \times 10^5\ \text{J mole}^{-1}$. By analogy with the equations characterizing the atomic theory of diffusion, ΔH can be regarded as the change in enthalpy associated with the diffusion of potassium by intercalation in the glassy carbon.

A general conclusion may be made that intercalation by alkali metals is possible in all carbons and disrupts glassy carbon, presumably, because no plasticity is available at room temperature to accommodate intercalation stresses.

7.4   Dissolution in metals

At high temperatures all carbide-forming metals attack glassy carbon and carbon fibres. This is of great importance to the development of carbon-fibre-reinforced metals and so will be dealt with at some length.

Jackson (1969) studied the compatibility of various carbon fibre–metal matrix combinations and identified two fibre-weakening mechanisms: formation of a surface carbide with Al and Cr, and structural recrystallization by Ni and Co. Platinum and copper were also found to seriously degrade the fibre even though non-reactive with graphite. Jackson and Marjoram (1970) found that carbon fibres in a Ni or Co matrix were recrystallized to large graphite particles at 1100 °C. They suggested that the mechanism for this process was analogous to an activated solid-state sintering process whereby the carbon atoms dissolved in the Ni and diffused rapidly to preferred sites where they nucleated and grew into a more ideal graphitic form. Their activation energy data indicated that the rate-controlling step was the carbon diffusion through the nickel. Interestingly, high-modulus fibres were more resistant to recrystallization than carbonized fibres, since the carbon atoms in the more stable form were less likely to become detached and dissolve in the nickel.

Metallic additions were seen to catalyse graphitization in non-graphitizing carbons during heat-treatment between 1400 and 2300 °C (Yokokawa *et al.*, 1966). Cobalt, nickel, copper, aluminium and manganese produced noticeable graphitization at 1500 °C. Yokokawa *et al.* showed that graphitization was probably catalysed by the formation of intermediate complexes of carbon and metals.

Fitzer and Kegel (1968) investigated the high temperature surface reactions between molten carbides (of V, Ti, Zr, Ni and Fe) and glassy carbon, pyrolytic carbons and natural graphite. They showed that glassy carbon was rapidly attacked by a vanadium carbide melt, and observed a deep interaction zone of well-ordered graphite and carbide particles. Pyrolytic carbon was attacked less rapidly, while natural graphite exhibited no interactions. The authors proposed that the VC wets and penetrates the carbon and dissolves carbon atoms. The molten carbide will then be saturated with respect to disordered carbon and supersaturated with respect to ordered carbon. The carbon atoms will precipitate from solution in the form of graphite. The driving force for graphitization is the difference in free energy between disordered carbon and highly ordered graphite; consequently, the glassy carbon interaction would be greater than pyrolytic carbon or natural graphite interactions.

Catalytic graphitization by iron and ferro-silicon was effectively explained by Baranieke *et al.* (1969). They found that the addition of silicon to iron accelerates graphitization and lowers the graphitization temperatures from 2500–3000 °C to approximately 1500 °C. The following graphitization mechanism was proposed: at 1600 °C molten ferro-silicon in contact with carbon will be saturated with respect to carbon but supersaturated with respect to graphite, which precipitates out, allowing dissolution of more carbon which again reprecipitates as graphite. The basic mechanism of graphitization depends on the transport of carbon atoms through the liquid metal, which in turn depends on the solubility and diffusion rate of carbon in iron. The addition of silicon tends to increase the diffusion rate of carbon atoms, and hexagonal SiC may act as nucleation sites for the graphite. On the other hand, the solubility of carbon in iron decreases with increasing silicon content; therefore, there is an intermediate optimum silicon content. Microstructural observations indicate that as graphitization proceeds, graphite crystals tend to grow through the molten ferro-silicon particles, splitting them up and moving them into contact with original carbon, allowing the process to repeat itself. The size of the graphite crystallites formed decreases as the ferro-silicon particles decrease in size. The final micro-structure is a fine submicron dispersion of ferro-silicon particles in a graphite matrix.

The processes discussed above are largely dependent on the surface structure and properties of the catalyst (metal or carbide) and the carbon phase, and the corresponding interfacial relationships. When a molten phase comes into contact with a solid phase, the results are usually described by wetting theory. Very little information is available about the nature of wetting of graphites by liquid metals. Munson (1967) determined the surface energies of some liquid metal interfaces with carbon, and Buhsmer and Heintz (1969) applied classical wetting theory to some non-reactive liquid metal–graphite systems. Wetting is most important in the production of carbon-fibre-reinforced metal composites by infiltration of the liquid metal. Pure copper is attractive since it is inert to carbon and graphite. However, for the same reason, C–Cu composites prepared by liquid infiltration are not significantly strength-ened since no interfacial bond is formed. Mortimer and Nicholas (1970) found that small additions of chromium and vanadium help promote wetting by forming a thin adherent layer of carbide at the graphite/metal interface. They found no correlation between chemical reactivity and wetting behaviour, and suggested that the surface properties of the carbides formed determined the degree of wetting.

At low (up to 1000 °C) temperatures polymeric carbons are inert to metal attack apart, of course, from alkali metals. However, at high temperatures in our laboratory, we have found that the carbon surface is easily wetted (Matthews, 1970). A typical example is as follows. Molybdenum metal powder was placed on a glassy carbon sheet and on a graphite sheet and taken up to 2700 °C in an argon-filled carbon resistance furnace. On a perfect graphite sheet the molybdenum is trans-formed to Mo_2C and moves across the graphite surface eating up car-bon atoms in the upper sheets only. Figure 45 shows the molybdenum and carbide/glassy carbon interface. The melting angle is virtually zero

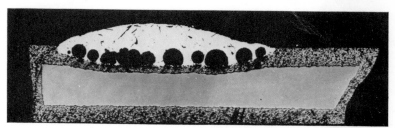

Fig. 45. Attack of molybdenum at high temperatures on a glassy carbon disc.

and a fine dispersion of molybdenum carbide/recrystallized carbon has completely surrounded the surface of the carbon disc via a dissolution precipitation mechanism. Higher magnification shows the fine carbide dispersion in recrystallized carbon and a decreasing particle size with depth of penetration. The interaction obviously proceeds more rapidly at the carbon surface.

Provided the process temperature is not too high, glassy carbon can be incorporated in molten aluminium with no apparent chemical reaction, thus establishing a new range of composite materials (Ryan, Kawamura and Jenkins, 1969).

8 Applications and uses of polymeric carbons

We have shown that carbon fibres, chars and glasses all belong to the same group of materials which we have termed Polymeric Carbons. The common structure is of various arrangements of carbon ribbons forming a network of microfibrils. The physical, mechanical and chemical behaviour are all directly related to this structure. There only remains to be added a brief indication of the uses and applications of carbon glasses and fibres.

The uses and applications of activated carbons were mentioned in the previous chapter. A more detailed account can be found in the article of F. A. P. Maggs (1968).

8.1 Applications of carbon glasses

The outstanding qualities of glass-like carbon are:

Chemical inertness and mirror-like finish;
Reasonable electrical conductivity;
Hardness in isotropic state;
Impermeability to gases and liquids;
Ability to withstand high temperatures in a non-oxidizing atmosphere and good thermal shock resistance.

The main difficulties in applying the materials are:

High expense due to technological difficulties in manufacture;
Inevitable weight loss and expensive raw materials;
Difficulties in machining glassy carbon, keeping to a close tolerance;
Impossibility of making non-porous glassy carbon artefacts thicker than 0.3 cm;
Brittleness of isotropic material; poor resistance to mechanical shock.

With these points in mind we shall examine present and future applications.

A typical range of glassy carbon ware is illustrated in fig. 46 by courtesy of the Tokai Electrode Company of Japan.

The rapid growth of integrated circuits needing epitaxial growth on silicon crystals led to the first large-scale use in Japan of glass-like carbon as a susceptor (Yamada, 1968). It has proved very successful, especially following fundamental improvements in the forming process.

Fig. 46. Examples of glassy carbon ware. (Courtesy of S. Yamada.)

The main requirements are a low impurity content and a low effective surface area. Large discs with diameters of between 30 and 40 cm have been produced by Tokai for this purpose.

Another use, in parallel to the above, is for crucibles for the preparation and growth of single crystals. To prepare large crystals of CaF_2, for instance, cone-type crucibles with a volume of about 10 litres are used. The operation is carried out at 1400 °C at 10^{-5} torr. The discharging process is greatly simplified because of the chemical inertness of the carbon which promotes ease of separation. After-treatments such as the remelting of the surface of the crystal are not found to be necessary.

Because of its mirror-like finish, impermeability and resistance to chemical attack, glassy carbon found early applications in the laboratory, for the manufacture of beakers, crucibles and dishes. Crucibles were used in processing GaP where previously quartz had been used. The non-ionic, non-porous properties of the carbon gave it preference when processing such high purity materials. It replaces platinum in the processing of CaF_2 and LaF_3. Chemical analyses involving fusions and acid-leaching are now carried out in glassy carbon crucibles in preference to platinum.

Another shape of crucible having dimensions $30 \times 16 \times 500$ mm has been introduced as a reaction vessel for crystallization of CdS and ZnS – a process which requires 1500 °C and a pressure of 50 atm. The vessel material does not contaminate and has a long life even under these extreme conditions.

Yamada (1968) reports that auxiliary electrodes have been used for laser-microprobe analysis because of their surface stability, homogeneity and impermeability. Le Carbone Lorraine also report the application of 'Carbone Vitreux' to the quantitative determination of lead and for spectrochemical analysis. Petrologists use sintered glassy carbon filters when using concentrated hydrofluoric acid in the analysis of strata.

Vitreous carbon now replaces PTFE in the manufacture of non-sticking burette taps because it does not tend to flow under pressure. Glassy carbon tapered joints compatible with glass equipment have been made for the handling of corrosive fluorides and alkalis.

Glassy carbon is not wetted by molten aluminium and so is an ideal material for drip pipes for dehydrogenation of molten ingots with chlorine. Crucibles and pipes prepared from glassy carbon are inert to attack in molten-salt electrolysis and so do not contaminate the end product.

Owing to its mechanical weakness and the impossibility of producing adequate wall thicknesses glassy carbon can only be used as a liner in large graphite vessels.

Heaters for electronic devices of all shapes and sizes have been produced because of glassy carbon's conducting properties; it is used for making slit plates for the electron beam of a linear accelerator. Voltages as high as 6×10^6 V are possible because resultant gamma and X-ray radiation is low; no melting occurs and no outgassing is necessary so that a vacuum of 10^{-6} torr is possible. Glassy carbon has been used for heating elements in coffee pots, and to replace the ceramic plates to improve heat transfer between cooker heating elements and cooking utensils.

Glassy carbon has been used as a brush for carbon track potentiometers because of its low electrical resistance, low noise, smooth surface and resistance to corrosion. It is also suggested for lightning protectors for telephone circuits, for microphone electrodes and relay contacts. Potentiometer brushes, like those used in radio volume controls, are normally made of carbon black with a suitable binder. These brushes move across a carbon track, reinforced by glass and resin binder. The track is abrasive and wears away the brushes; carbon dust then changes the resistance. Glassy carbon brushes, tested over a period equivalent to several times the normal life of these components have proved satisfactory; they do not wear and the resistance value of the track is unchanged.

The material's low coefficient of friction has also suggested research into rotating contacts for slip rings. The use of glassy carbon could lead to lower starting torques and lower friction when running on stainless steel.

Glassy carbon's ability to conduct electricity makes possible its use in induction-heated chemical plant and laboratory equipment. Because the material is impermeable it is suitable for the manufacture of heat exchangers – tubes will seldom scale up and, even if this does happen under certain conditions, the deposits can be removed chemically. Thermal conductivity of the material falls between that of glasses and carbons; heat transfer can be improved by using the stronger glassy carbon material with a thinner sectional thickness.

Prevention of static build-up is useful in the synthetic-fibre and glass-fibre industries. The fibres are fed through filament guides normally made of glass or alumina ceramic. Ceramic guides do not wear uniformly and they have to be polished after manufacture. Glassy carbon, on the other hand, has a high finish with a low coefficient of friction, and wear would be uniform on filament guides made of this material.

The reflective properties of the mirror-like finish of glassy carbon can be utilized in the manufacture of reflectors for high intensity lamps. Glass cannot be used because of the heat built up by the lamp, and the practice is to use polished nickel or tungsten which is hard to fabricate and costly. The black surface of glassy carbon is ideal for dissipating heat and only the minimum surface polishing is necessary.

Typewriter characters are another outlet. Eventually traditional materials burr over and oxides form, so that the imprint becomes blurred. Nothing sticks to glassy carbon and it does not burr.

Because of its non-sticking properties glassy carbon makes an excellent jig material. A glassy carbon mandrel has advantages in the manufacture of glass joints; it has the same hardness as glass and therefore does not wear as rapidly as a graphite mandrel. It is used as a jig material in the manufacture of glass-to-metal seals for transistors. Glass powder is poured into the jig, a metal strip inserted into the powder and the whole assembly passed through a furnace to complete the seal. This principle has been extended to the manufacture of flat packs.

Although vitreous carbon is an inert material and highly impermeable, living animal tissue will adhere to it. This suggests opportunities for the material in medical implants. In the USA a glassy carbon ring containing a biocompatible quartz window was implanted in a man's arm and healed completely in seven days without rejection; the window was still in place after six months. The skin sealed to the ring and no irritation, infection or rejection occurred. This has never happened in similar experiments with any other material, possibly because other materials absorb certain fluids from the body. Graphite, for example, stops the free movement of blood around the outside of the component and causes ulceration. Glassy carbon may well soon be used for components implanted inside the body, such as pins for broken bones, heart valves, additional tract to the liver, and tooth keys.

In the Rancho Los Amigos Hospital, Los Angeles 'biocarbon' has been used in amputation cases to make a cuff or sleeve to protect the passage of a longitudinal metal implant thrust into the medulla of the residual bone. Thus a fore-arm prosthesis, for instance, is successfully fixed directly to the bone of the stump, a fixation considered hitherto impossible.

8.2   Applications of carbon fibre

Cloth and yarn consisting of low modulus carbon fibres derived directly from Rayon have been marketed since the late fifties (Cranch, 1961). Cloth, cord, felt, wool and sponge of cellulose, PAN, or any of the other possible precursors for polymeric carbon can be carbonized simply and directly to carbon cloth, cord, felt, wool and sponge, as illustrated in fig. 47. Such materials are manufactured for use as filters, catalyst carriers and thermal insulation.

Homsy (1974) has used carbon fibre together with a little PTFE to coat prosthetic implants. The coating acts as a porous low modulus interface for the stabilization of soft tissue and bone prostheses. Tissue ingrowth is excellent and the interface accommodates movements in the implanted prostheses with respect to lost tissue and bone, thus obviating planes of unwanted stress discontinuity.

The important property of polymeric carbons is the possibility of orienting the microfibrils along a fibre axis. This has resulted in fibres with high specific modulus and strength which are eminently suitable for reinforcing various matrices. The most successful use has been in reinforcing plastics.

A comparison between the various fibres used for fibre reinforcement is presented in table 12. It should be noted that carbon has close com-

TABLE 12   *Comparison of commercial fibres used for reinforcement*

Fibre	du Pont PRD/49	Boron	E-glass	HTS glass	Carbon fibre Type I	Carbon fibre Type II
Specific gravity	1.47	2.56	2.5	2.5	1.9	1.75
Modulus (GNm^{-2})	140	400	40	80	400	260
Strength (MNm^{-2})	2800	3300	2000	2700	2000	3000

Fig. 47. Carbon fibre felt and cord. (Courtesy of Conradty.)

petition from the du Pont synthetic organic fibre, PRD/49, and boron. All the strengths are in the same range (2–3000 $MN m^{-2}$). It is only in stiffness that carbon fibre shows improvement over the much cheaper commercial E-glass.

8.2.1 Carbon-fibre-reinforced plastics (CFRP)

An excellent review of this use has been presented by Gunston (1969); more recently, a useful book on CFRP has been written by Gill (1972). To make a composite, several routes may be followed. The simplest is the leaky mould technique. A mould to fit the part is made in two halves and the lower section filled with an excess of hot-setting phenolic or epoxy resin. Enough wetted fibres are then laid in the mould to fill the moulding cavity when closed, with their orientation arranged in the direction requiring maximum strength. The upper section of the mould is brought down, squeezing out the excess resin. The component is then cured, typically for some hours at about 100 °C and usually under some pressure. The finished composite, which is generally also the finished part, requires only superficial trimming at the ends and removal of any 'flash' which may project along the line where the two portions of the mould joined.

This method is suitable for making all kinds of simple shapes, and is capable of being automated if bulk production of similar parts is called for; but it is not appropriate to complicated forms, which have to be made in either of two ways, of which the more challenging is filament-winding. This involves the creation of a three-dimensional shape by winding a continuous length of wetted fibre round and round it. A sphere is wound like a ball of string; almost any other shape involves considerable calculation if the orientation of the filament at each point is to be that best suited to bearing the loads the component will experience in use. In the USA various filaments have been used to wind very large cases for solid-fuel rocket motors, and automated filament-winding has in consequence reached an advanced stage. Inherent difficulties make its adoption slow.

Far easier is the 'prepreg' system which produces the composite in the form of a raw material which is only later fashioned into the desired shape. The raw material is made in the form of uniform sheet or tape known as 'warp, 'oriented fibre' or 'prepreg'. Prepreg is made by dipping numerous tows or groups of fibres in a dilute solution of resin in acetone and then laying them down side by side, exactly parallel and without overlap, on a firm flat surface. The latter should not react with the fibres in any way; Teflon film or siliconized paper has been found suitable. A second sheet of this material is placed on top; the assembly is warm-rolled to even thickness, eliminating voids between fibres, and dried in the usual way. The solvent evaporates and the resin binds the fibres into sheets with a precise thickness down to about 0.02 mm; tape is used by uncoiling from a spool and stripping off the facing films. Prepreg can be cut like card. All carbon fibres in bulk use are marketed in prepreg forms, except in special cases where users have a component that must be filament-wound.

There are yet other ways of using carbon fibres as reinforcement. One can make use of fibre scraps that might otherwise be wasted; such short lengths are chopped up and used to form a moulding compound with either thermoplastics or thermosetting resins, often with remarkably beneficial results.

Carbon composites are especially well suited to making things are either slender or have the form of a two-dimensional sheet rather than a solid block. Their design is more difficult if they are subjected to complex systems of loading, but the stresses imposed on fan blades are not complicated and these parts are ideally suited to this new material. A blade's profile and angle of twist vary continuously from root to tip. The dominant loading is the centrifugal pull from root to tip due to the weight of the blade itself. Superimposed on this is a bending load caused by the work done by the blade on the airflow.

Such a blade could be made as a single composite moulding, or it could, with difficulty, be wound from continuous filament. Fabricators actually cut a number of laminates of different shapes and sizes from prepreg sheet and bond these together in a single moulding process. The system of loading calls for most of the laminates to have their fibres running along the length of the blade. The aerofoil section of the blade could be built up by incorporating the smaller laminations, not on the outer faces but inside the blade, so that the outer plies cover the whole area. But a blade made wholly of plies with fibres running from root to tip would twist easily. Accordingly, each side of the blade incorporates two laminations with fibres orientated at 45° to left and right.

The shape of each lamination is calculated on a computer and then slightly modified. Each particular laminate is cut in multiples by a 'pastry cutter' at the correct orientation. The operator then takes all the plies needed to make a blade, one at a time and in the correct order, and places them in sequence on a template having the same camber and twist as the finished blade and scribed with the precise outline of each ply. One half of the blade is built up by inserting glass-fibre packing strips at the root to open it out to the desired double wedge form. The two halves are then brought face to face and the assembly hot-pressed in a mould. Excess resin is squeezed out and, after curing, the blade is post-cured at a suitable temperature, and 'flash' trimmed off, and the root precision ground. An example of such a fan blade assembly made out of CFRP is shown in fig. 48.

Outside the purely commercial world the most obvious application is spacecraft. It costs so much to put a kilogramme into deep space that every space payload is studied for possible incorporation of carbon composites. The main load-carrying structure of a satellite is commonly a light alloy sandwich but test structures have been made with facing skins replaced by two plies of 0.25 mm prepreg sheet with fibres running vertically in one and horizontally in the other.

There are two fundamental problems associated with carbon composites that demand further development. Firstly, the stiffness is so great that techniques traditional in metal workshops are no longer

Fig. 48. Carbon fibre reinforced resin blades used in a variable pitch fan. (Courtesy of Dowty.)

appropriate. Suppose a part is to be fixed at both ends to a rigid frame. With metal, one end can be fastened and the other then slightly pushed or pulled until the bolt or rivet can be inserted. With carbon composite no significant deflection can be imposed; the load and stress become enormous and the part fails before it can be seen to deform. The second basic difficulty is that of joints, especially of joints between composite and metal.

With thermoplastic and carbon fibres certain difficulties are experienced in adhesion, fibre orientation and filament breakage due to high shear forces when moulding. When the fibre–matrix bond is weak, the strength and modulus of one or both constituents individually are ineffective. In the case of a composite using a thermosetting resin the problem has been overcome by processing the carbon fibre surface to improve the bonding between fibre and matrix. The strength of the bond between fibre and matrix is proportional to fibre length and so when short fibres are used the strength of this bond becomes critical under shear conditions. In a moulding operation the fibres are subject to high shear forces and because of their high modulus the fibres are liable to break, thus giving rise to a lower bond strength between fibre and ma because of the shorter lengths of fibre. Nevertheless, the modulus of a thermoplastic composite is invariably increased and a higher specific modulus is obtained for the composite by using carbon fibres in preference to other types of reinforcing fibres such as glass or asbestos. The increase in specific modulus is not so great as with thermosetting composites and this is why the initial interest in carbon fibre thermoplastic composites is for applications requiring a high modulus together with another quality, for instance, hard wear, the classic example of this being the gear-wheel.

Thus as a lightweight structural material, carbon fibres have a unique place because of their high modulus:density ratio. Additionally, the use of carbon fibres will often be justified after consideration of a number of factors together, such as strength, self-lubrication, resistance to wear, electrical conduction, low magnetic response, low density.

Resins are ideally suited to binding together carbon fibres, but epoxies are limited to use at about 200 °C. Polyimides are capable of raising the limit to 350 °C or slightly higher; polyphenylenes also extend the working range to 300 to 400 °C. Rogers and Kingston-Lee (1971) have used carbon fibres to reinforce a new polyimide laminating resin which is stable and readily soluble. A unidirectional fibre laminate possessed flexural strength of $800 \, MN \, m^{-2}$. Even after 1774 hours at 300 °C it retained a strength of $336 \, MN \, m^{-2}$. It will be possible to use such composites for aerospace structures, hot-air ducting, fan blades, oven fittings and engine components.

8.2.2   Carbon-fibre-reinforced metals and ceramics

Carbon fibres bonded in a refractory matrix of metal or ceramic enable a new order of light, stiff and strong construction to extend into high temperature fields such as gas turbine blading, nuclear and plasma devices, rocket motors, and various other types of system which cannot yet be built at all. Use of a metal matrix would naturally make for a stronger composite at all temperatures, especially at above 200 °C, and would greatly improve its conductivity of electricity and heat. But the carbon fibres tend to go into solution in the metal, and with many alloys there is a chemical reaction yielding a carbide. Extensive research suggests that the best answer may be to protect each fibre within a barrier layer of some material having almost the same strength, elastic modulus and coefficient of expansion as the carbon, yet totally unreactive with it. Alumina and some other oxides show promise in this role, but it is not proving easy to achieve barriers that grip both fibre and matrix and do not crack in severe and prolonged testing. If the environment never gets hotter than 600–800 °C it is possible to use a matrix of copper or tin and allow the inevitable oxide layer to act as a barrier to carbide formation.

8.2.3   Carbon-fibre-reinforced carbon (CFRC)

The production of CFRC is an obvious step subsequent to the development of carbon fibre. Theoretically, it is now possible to manufacture a material with all the advantages of pure carbon but stronger and stiffer than many steels at room temperature, with a fraction of their density. More significantly, CFRC will retain these good mechanical properties to very high temperatures, creep being negligible below 2000 °C. The uniqueness of this material would allow the growth of new technology, heretofore believed impossible.

The main problem is to introduce carbon into the interstices between fibres, without causing distortions and without introducing internal strains sufficient to disrupt the composite. There are basically two possible techniques.

Firstly, one can introduce a suitable resin such as phenolic resin and then carbonize the composite in order to convert the resinous component to a polymeric carbon *in situ*. The result is a polymeric carbon reinforced with a polymeric carbon fibre. The process is described by Mackay (1969).

The simple process is not satisfactory because the resins contract by up to 40% on carbonization, while the fibres retain their original dimensions. The resultant stresses are sufficient to disrupt the original resin–fibre bonds, to produce a weak, porous composite. The Atomic Weapons Research Establishment (Hill *et al.*, 1973) have demonstrated that this problem can be overcome by dispersing 20% by volume of colloidal graphite in the original resin. The presence of this inert finely divided component lowers the contraction of the resinous binder on carbonizing; the integrity of the composite is thereby preserved. For example, specimens made by a leaky mould technique using fibres laid uniaxially and subsequently heated to 2500 °C possess tensile strengths in excess of 700 MNm^{-2} and Young's modulus in excess of 120 GNm^{-2} in the direction parallel to the fibres' axis. The density of such material is only 1.5 gcm^{-3}.

The second process for filling interstices between fibres is that of Chemical Vapour Deposition (CVD). This is described in detail by Kotlensky (1973). Briefly, a hydrocarbon gas mixed with an inert carrier gas is allowed to pass through a porous carbon mass at temperatures high enough for one of the many forms of pyrolytic carbon to be de-

Fig. 49. The structure of carbon-fibre-reinforced carbon used for brake linings, × 4400. (Courtesy of Dunlop.)

posited. The result is a carbon-fibre-reinforced pyrolytic carbon, as illustrated in the electron micrograph in fig. 49. This process has met with success in the USA (Kotlensky, 1969; McLoughlin, 1970).

In the UK, a moulded carbon technique has been developed successfully by Fordath Ltd, based on the original work at the Royal Aircraft Establishment (Bickerdike *et al.*, 1967). Carbon fibres are impregnated with a small amount of phenolic resin and pressure moulded to a desired shape. The moulding is cured and carbonized. Subsequently, the many pores are filled by pyrolytic deposition from a suitable hydrocarbon gas. The process lends itself to high dimensional accuracy and excellent conformation to the original mould. Such mouldings have been made with cheap, isotropic fibre in the form of cloth or felt. In this way isotropic composites can be made with strength and stiffness superior to finest isotropic electrographites.

Fig. 50. Aircraft brakes of carbon-fibre-reinforced carbon. (Courtesy of Dunlop.)

The development of CFRC has been accelerated by the demands of advanced aerospace technology (Vaccari, 1969; Parmee, 1971). The immediate use is for re-entry vehicle heat shields, temperature ducting systems, nuclear rocket engines, hot-pressing dies, bearing materials, and rocket nozzles. Dunlop have adopted CFRC for lightweight brakes suitable for heavy, fast jet aircraft; fig. 50 illustrates this.

The unique biocompatibility of carbon combined with the excellent

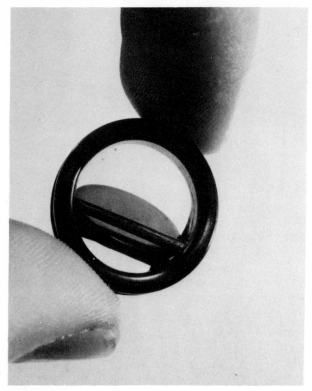

Fig. 51. Heart valve made entirely out of carbon using carbon-fibre-reinforced carbon designed and fabricated in Brazil.

mechanical properties of CFRC make it suitable for all kinds of medical implant in living tissue – even for percutaneous limb extensions. Work is progressing in Swansea and at the Centro Tecnico Aerospacial, Brazil, directed to using CFRC in medical prostheses. CFRC rods made in Brazil by a variation of the resin impregnation technique are used instead of Steinman pins made of stainless steel in bone extensions. Figure 51 shows a heart valve designed and fabricated in Brazil, using CFRC, for the Escola Paulista de Medecina – it is the first heart valve to be made entirely of carbon.

We may appropriately conclude by emphasizing that CFRC materials consist wholly of carbon and, mainly, of polymeric carbon. They are the strongest and stiffest engineering materials that can be made wholly of carbon. It is by this new technology of tailoring material by suitable arrangement of carbon fibrils derived from polymeric precursors that we can achieve the ultimate in mechanical properties.

References

General references

Porous Carbon Solids, edited by R. L. Bond, Academic Press, London and New York, 1967

Les Carbones, Groupe Française d'Etude des Carbones, Masson et Cie, Paris, 1965

Activated Carbon, J. W. Hassler, Leonard Hill, 1967

Chemistry and Physics of Carbon, series edited by P. L. Walker jun., Dekker, New York:

    M. M. Dubinin: *Porous structure and adsorption properties of active carbons*, in Vol. 2, 1966

    Sabri Ergun: *X-ray studies in carbon*, in Vol. 3, 1968

    W. Ruland: *X-ray diffraction studies on carbon and graphite*, in Vol. 4, 1968

    E. Fitzer, K. Müller and W. Schäfer: *The chemistry of the pyrolytic conversion of organic compounds to carbon*, in Vol. 7, 1969

    R. Bacon: *Carbon fibres from Rayon precursors*, in Vol. 9, 1973

    W. V. Kotlensky: *Deposition of pyrolytic carbon in porous solids*, in Vol. 9, 1973

    W. N. Reynolds: *Structure and physical properties of carbon fibres*, in Vol. 11, 1973

    G. M. Jenkins: *Deformation mechanisms in carbons*, in Vol. 11, 1973

Graphite and its Crystal Compounds, A. R. Ubbelohde and F. A. Lewis, Oxford U.P., 1960

Manufactured Carbon, edited by H. W. Davidson, Pergamon, New York, 1968

    F. A. P. Maggs: Chapter 4, *Activated charcoal*

Detailed references

Adams, L. B., E. A. Boucher, R. N. Cooper and D. H. Everett (1970) in *Proceedings of the Third Conference on Industrial Carbon and Graphite*, Society of Chemical Industry, London

Adkins, C. J., S. M. Freake and E. M. Hamilton (1970) *Phil. Mag.* **22**, 183

Andrew, J. F. and S. Sato (1964) *Carbon* **1**, 225

Antonowicz, K., L. Cacho and J. Tunlo (1973) *Carbon* **11**, 1

Arévalo, M. C. and G. M. Jenkins (1974) in *Proceedings of the Conference on Industrial Carbon and Graphite*, Society of Chemical Industry, London, p. 116

Bacon, G. E. (1956) *J. Appl. Chem.* **6**, 477

Bacon, R. (1960) *J. Appl. Phys.* **31**, 284

Bacon, R. and W. A. Schalamon (1967) in *Proceedings of the Eighth Conference on Carbon, Buffalo*

Bacon, R. and W. A. Schalamon (1969) *Appl. Polymer Symp.* No. 9, Amer. Chem. Soc., p. 285

Bacon, R. and W. H. Smith (1965) in *Proceedings of the Second Conference on Industrial Carbon and Graphite*, Society of Chemical Industry, London

Bacon, R. and M. M. Tang (1964) *Carbon* **2**, 227

Badami, D. V. and R. J. Hussey (1963) *Proceedings of the Sixth Conference on Carbon, Pittsburgh*, p. 72

Badami, D. V., J. C. Joiner and G. A. Jones (1967) *Nature* **215**, 386

Bale, E. S. (1970) *Proceedings of the Third Conference on Industrial Carbons and Graphite*, Society of Chemical Industry, London, p. 23

Baranieke, C., P. H. Pinchbeck and F. B. Pickering (1969) *Carbon* **7**, 213

Berg, C. A., E. H. Cumpston and A. Rinsky (1971) *Fibre Science and Technology* **4**, 153

Bickerdike, R. L., G. Hughes, E. Parkes and F. J. Robinson (1967) *Foundry Trade Journal* **38**, 3

Blackslee, O. L., D. G. Proctor, E. J. Seldin, G. B. Spence and T. Weng (1970) *J. Appl. Phys.* **41**, 3373

Blayden, H. E. and D. T. Westcott (1961) *Proceedings of the Fifth Conference on Carbon*, Vol. 2, Pergamon, New York, p. 97

Boucher, E. A., R. N. Cooper and D. H. Everett (1970) *Carbon* **8**, 597

Bradshaw, W. and P. Pinoli (1967) in *Proceedings of the Eighth Conference on Carbon, Buffalo*

Brydges, W. T., D. V. Badami, J. C. Joiner and C. A. Jones (1969) *Appl. Polym. Symp.* No. 9, American Chemical Society, p. 225

Bücker, W. (1973) *J. of Non-Crystalline Solids* **12**, 115

Buhsmer, C. P. and E. A. Heintz (1969) *J. Mats. Science* **4**, 592

Butler, B. L. (1973) *Sandia Laboratories Report* SLA-73-0485

Butler, B. L. and R. J. Diefendorf (1969) *Proceedings of the Ninth Conference on Carbon, Boston*, p. 161

Conor, P. N. and C. N. Owston (1969) *Nature* **223**, 1146

Cooper, G. A. and R. M. Mayer (1971) *J. Mats. Sci.* **6**, 60

Cranch, G. E. (1961) in *Proceedings of the Conference on Carbon*, Pergamon, New York, p. 589

Curtis, G. J., J. M. Milon and W. N. Reynolds (1968) *Nature* **218**, 83

Dacey, J. R. and D. A. Cadenhead (1960) in *Proceedings of the Fourth Conference on Carbon, Buffalo*, Pergamon Press, New York, p. 315

Dacey, J. R. and D. G. Thomas (1954) *Trans. Faraday Soc.* **54**, 250

Davidson, H. W. (1962) *Nuclear Engineering* **7**, 159

Davidson, H. W. and H. H. W. Losty (1963) *GEC Journal* **30**, 22–30

Diamond, R. (1959) *Phil. Trans. A* **252**, 193

Diefendorf, R. J. (1972) *Composite Materials*, AGARD Lecture Series, NATO

Easton, A. and Jenkins, G. M. (1974) in *Proceedings of the Conference on Industrial Carbon and Graphite*, Society of Chemical Industry, London, p. 79

Edstrom, T. and I. C. Lewis (1969) *Carbon* **7**, 85–91

Eley, D. D. (1967) *J. Polym. Science, C* **17**, 73

Ergun, S. (1969) in *Proceedings of the Ninth Conference on Carbon, Boston*

Ergun, S. and L. E. Alexander (1962) *Nature* **195**, 765

Ergun, S. and V. H. Tiensuu (1959) *Nature* **183**, 1668

Everett, D. H. and E. Redman (1963) *Proc. Chem. Soc.* **91**

Ezekiel, H. M. (1970) *J. Appl. Phys.* **41**, 5351

Ezekiel, H. M. (1971) *U.S. Air Force Material Laboratory Report* AFML-TR-69-167

Fischbach, D. B. (1967) *Carbon* **5**, 565

Fischbach, D. B. (1969) *Carbon* **7**, 196

Fischbach, D. B. and W. V. Kotlensky (1965) *NASA Technical Report* 32, 842

Fitzer, E. and B. Kegel (1968) *Carbon* **6**, 433

Fitzer, E. and W. Schäfer (1970) *Carbon* **8**, 597

Fitzer, E., W. Schäfer and S. Yamada (1969) *Carbon* **7**, 643

Fitzer, E., K. Müller and W. Schäfer (1969a) in *Chemistry and Physics of Carbon*, Vol. 7, Dekker, New York.

Fourdeux, A., C. Hérinckx, R. Perrett and W. Ruland (1969) *C.R. Acad. Sci. Paris, C* **269**, 1597

Franklin, R. E. (1951) *Proc. Roy. Soc. London A* **209**, 196

Friedman, H. L. (1963) *J. Polym. Science, C, Polymer Symposium*

Furukawa, K. (1964) *J. Cryst. Japan* 6, 101

Gay, R. and H. Gasparoux (1965) *Les Carbones*, Vol. 1, p. 63

Gilbert, J. B. and J. J. Kipling (1962) *Fuel* (London) 41, 249

Gill, R. M. (1972) *Carbon Fibres in Composite Materials*, Iliffe, London (on behalf of the Plastics Institute)

Griffith, A. H. (1920) *Phil. Trans. Roy. Soc. A* 221, 163

Guinier, A. and G. Fournet (1955) *Small Angle Scattering of X-rays*, Wiley, New York

Gunston, W. T. (1969) *Science* 5, 39

Gutmann, F. (1967) *J. Polym. Sci., C* 17, 41

Gutmann, F. and L. E. Lyons (1967) *Organic Semiconductors*, Wiley, New York

Halpin, M. K. and G. M. Jenkins (1968) *Nature* 218, 950

Halpin, M. K. and G. M. Jenkins (1969) *Proc. Roy. Soc. London A* 313, 421

Hawthorne, H. M., C. Baker, R. H. Bentall and K. R. Linger (1970) *Nature* 227, 946

Helberg, H. W. and B. Wartenberg (1970) *Phys. Stat. Sol.* 3, 401

Hérinckx, C. (1973) *Carbon* 11, 199

Hérinckx, C., R. Perret and W. Ruland (1968) *Nature* 230, 63

Hérinckx, C., R. Perret and W. Ruland (1972) *Carbon* 10, 711

Hill, J., E. J. Walker and C. R. Thomas (1973) in *Proceedings of the Eleventh Biennial Conference on Carbon, Tennessee*, Paper FC-19

Homsy, C. A. (1974) in *Cambridge Conference on Materials in Medicine*, Biological Engineering Soc.

Honda, T. (1966) *Tanso, Japan* 45, 19

Honda, T. and Y. Samada (1966) *Tanso, Japan* 46, 2

Inokuchi, H. (1951) *Bull. Chem. Soc. Japan* 24, 222

Inokuchi, H. and H. A. Kamatsu (1961) *Solid State Physics A.P.* 12, 93

Jackson, P. W. (1969) *Metals Engineering Quarterly* 9, 22

Jackson, P. W. and J. R. Marjoram (1970) *Nature* 218, 83; *J. Mats. Sci.* 5, 9

Jackson, P. W. and J. R. Marjoram (1970) *J. Mats. Sci.* 5, 9

Jenkins, G. M. (1973) in *Chemistry and Physics of Carbon*, Vol. 11, ed. P. L. Walker, Dekker, New York

Jenkins, G. M. and G. Jouquet (1968) *Carbon* 6, 85

Jenkins, G. M., K. Kawamura and L. Ban (1972) *Proc. Roy. Soc. London A* 327, 501

Johnson, D. J. (1971) International Conference on Carbon Fibres, the Plastics Institute, London, paper 8

Johnson, D. J. and C. N. Tyson (1970) *Brit. J. Appl. Phys. (J. Phys. D)* 3, 526

Johnson, J. W. (1969) *Appl. Polym. Symp.* No. 9, p. 229

Johnson, J. W., J. R. Marjoram and P. G. Rose (1969) *Nature*, 221, 357

Johnson, W. (1970) in *Proceedings of the Third Conference on Industrial Carbon and Graphite*, Society of Chemical Industry, London, p. 447

Johnson, W. and W. Watt (1967) *Nature*, 215, 384

Jones, P. F. and R. G. Duncan (1971) *J. Mats. Sci.* 6, 289

Jones, W. R. and J. W. Johnson (1971) *Carbon* 9, 645

Kakinoki, T. (1965) *Acta Cryst.* 18, 578

Kakinoki, T., K. Katada, T. Karawa and J. Iro (1960) *Acta Cryst.* 13, 171

Kawamura, K. (1971) Ph.D. Thesis, University of Wales

Kawamura, K. and G. M. Jenkins (1970) *Industrial Carbon and Graphite*, Society of Chemical Industry, London, p. 98

Kawamura, K. and G. M. Jenkins (1970) *J. Mats. Sci.* 5, 262

Kawamura, K. and G. M. Jenkins (1971) *Nature* 231, 175

Kawamura, K. and G. M. Jenkins (1972) *J. Mats. Sci.* 7, 1099

Kiive, P. and S. Mrozowski (1959) in *Proceedings of the Third Conference on Carbon*, Pergamon, p. 165

Kipling, J. J. and B. McEnaney (1966) in *Proceedings of the Second Conference on Industrial Carbon and Graphite*, Society of Chemical Industry, London, p. 380

Kipling, J. J., J. N. Shewood and P. V. Shooter (1964) *Carbon* **1**, 315
Klug, H. P. and L. E. Alexander (1954) *X-ray Diffraction Procedures*, John Wiley, New York
Kotlensky, W. V. (1969) in *Proceedings of the Ninth American Carbon Conference*, to be published in series *Chemistry and Physics of Carbon*, Dekker, New York
Krishnan, K. S. and N. Ganguli (1939) *Nature* **144**, 667
Kroger, C. and W. Ruland (1955) *Chem. Techn. Brenstoffe Veredlungs Produkte* **36**, 97
Lewis, J. C. (1965) in *Proceedings of the Second Conference on Industrial Carbon and Graphite*, Society of Chemical Industry, London, p. 258
Lewis, J. C., B. Redfern and F. B. Cowlard (1963) *Solid State Electronics* **6**, 251
Linger, K. R., H. W. Baker, R. H. Bentall and H. M. Hawthorne (1970) *Nature* **227**, 946
Loebner, E. E. (1955) *Phys. Rev.* **102**, 46
Losty, H. H. W. and H. D. Blakelock (1965) in *Proceedings of the Second Conference on Industrial Carbon and Graphite,* Society of Chemical Industry, London, p. 320
Mackay, H. A. (1969) *Sandia Labs. Report.* SC-RR-68-651
McLintock, I. S. and J. C. Orr (1973) in *Chemistry and Physics of Carbons*, Vol. 11, ed. P. L. Walker, Dekker, New York, p. 243
McLoughlin, J. R. (1970) *Nature* **227**, 701
Madorsky, S. L. (1953) *J. Res. Natl. Bur. Standards* **51**, 327
Madorsky, S. L. (1964) *Thermal Degradation of Organic Polymers*, Interscience, New York
Maggs, F. A. P. (1968) in *Manufactured Carbon*, Pergamon, New York
Marsh, D. M. (1964) *Proc. Roy. Soc. A* **282**, 33
Marsh, H. and W. F. K. Wynne-Jones (1964) *Carbon* **1**, 269
Matthews, B. (1970) Ph.D. Thesis, University of Wales
Metrot, A. and A. Hérold (1967) *C. r. hebd. Séanc. Acad. Sci. Paris* **264**, 8
Moreton, R. (1969) *Fibre Science Techn.* **1**, 273
Moreton, R. (1970) *Proceedings of the Third Conference on Industrial Carbon and Graphite*, Society of Chemical Industry, London, p. 320
Moreton, R., W. Watt and W. Johnson (1967) *Nature* **213**, 690
Mortimer, D. A. and M. Nicholas (1970) *J. Mats. Sci.* **5**, 149
Mott, N. F. and E. A. Davies (1971) *Electronic Processes in Non-Crystalline Materials*, Clarendon Press, Oxford
Mrozowski, S. (1952) *Phys. Rev.* **85**, 609; *Phys. Rev.* **86**, 1056
Mrozowski, S. (1956) in *Proceedings of the First and Second Conference on Carbon, Buffalo*, Waverly, Baltimore, p. 31
Mrozowski. S. (1971) *Carbon* **9**, 97
Mrozowski, S. and A. Chaberski (1956) *Phys. Rev.* **104**, 74
Munson, R. A. (1967) *Carbon* **5**, 471
Noda, T. and M. Inagaki (1963) *Carbon* **1**, 86
Noda, T. and M. Inagaki (1964) *Bull. Chem. Soc. Japan* **37**, 1534
Noda, T. and H. Kato (1965) *Carbon* **2**, 289
Noda, T., M. Inagaki and S. Yamada (1968) *Bull. Chem. Soc. Japan* **41**, 3023
Otani, S. (1965) *Carbon* **3**, 31
Otani, S. (1966) *Carbon* **4**, 425
Otani, S. (1967) *Tanso* **50**, 32
Ouchi, K. (1966) *Carbon* **4**, 59
Ouchi, K. and H. Honda (1955) *J. Chem. Soc. Japan* **76**, 154
Ouchi, K. and H. Honda (1956) *J. Chem. Soc. Japan* **77**, 147
Ouchi, K. and H. Honda (1959) *Fuel (London)* **38**, 429
Owston, C. N. (1970) *Brit. J. Appl. Phys. (J. Phys. D)* **3**, 1615
Pacault, A., A. Marchand, P. Bothorel, J. Zanchetta, F. Boy, J. Cherville and M. Oberlin (1960) *J. Chim. Phys.* **57**, 892

Parmee, A. C. (1971) *Carbon Fibres*, Plastics Institute, London, Paper 38, p. 154

Perret, R. and W. Ruland (1969) *J. Appl. Cryst.* **2**, 209

Phillips, L. N. (1967) *Trans. J. Plastics Inst.* 589

Platzer-Rideau, N. (1960) *Annals. Chim.* **805**

Pohl, H. A. (1962) *Modern Aspects of the Vitreous State*, Vol. 2, Butterworths, London, p. 82

Reynolds, W. N. (1970) *Proceedings of the Third Conference on Industrial Carbon and Graphite*, Society of Chemical Industry, London, p. 427

Rogers, K. F. and D. M. Kingston-Lee (1971) *Carbon Fibres: Plastics Institute Conference*, Paper 34, p. 273

Rosenberg, B., B. B. Bhowmik, H. C. Harder and E. Postow (1968) *J. Chem. Phys.* **49**, 4108

Rüdorff, W. and E. Schulze (1954) *Angew. Chem.* **66**, 305

Ruland, W. (1965) *Carbon* **2**, 365

Ruland, W. (1967) *J. Appl. Phys.* **38**, 3585

Ruland, W. (1969) in *Chemistry and Physics of Carbon*, Vol. 4, ed. P. L. Walker, Dekker, New York

Ruland, W. (1969a) *Polymer* **9**, 1368

Ruland, W. (1969b) in *Proceedings of the Ninth Conference on Carbon, Boston*, Paper MP 24 (DCIC)

Ruland, W. and R. Perret (1968) *J. Appl. Cryst.* **1**, 308

Ryan, T., K. Kawamura and G. M. Jenkins (1969) in *Proceedings of the Ninth Conference on Carbon, Boston* (DCIC)

Schuyer, J. and D. W. Van Krevelen (1955) *Fuel (London)* **34**, 213

Sharp, J. V. and S. G. Burnay (1971) *Conference on Carbon Fibres*, Plastics Institute, London, Paper 10

Shindo, A. (1961) *J. Ceram. Ass. Japan* **69**, 195

Shindo, A. (1964) *Carbon* **1**, 391

Short, M. A. and P. L. Walker (1963) *Carbon* **1**, 3

Singer, L. S. (1968) in *Symposium on Carbonisation and Graphitisation*, Paris

Spain, I. L. (1973) in *Chemistry and Physics of Carbons*, Vol. 8, ed. P. L. Walker, Dekker, New York, p. 1

Standage, A. E. and R. Prescott (1966) *Nature* **211**, 169

Strauss, H. E. (1963) *Proceedings of the Fifth Conference on Carbon*, Vol. 2, Pergamon, p. 647

Takahashi, E. and W. Westrum (1970) *Proceedings of Tokyo Carbon Conference*

Takezawa, T. and G. M. Jenkins (1974) in *Proceedings of the Conference on Industrial Carbon and Graphite*, Society of Chemical Industry, London, p. 194

Tang, M. M. and R. Bacon (1964) *Carbon* **2**, 211

Taylor, R. E. and D. E. Kline (1967) *Carbon* **5**, 607

Tsuzuku, T. (1960) *J. Phys. Soc. Japan* **15**, 1373

Tsuzuku, T. (1964) *Carbon* **1**, 511

Tsuzuku, T. and H. Kobayashi (1961) in *Proceedings of the Fifth Conference on Carbon*, Vol. 2, Pergamon, p. 259.

Tsuzuku, T. and K. Saito (1966) *J. Appl. Phys (Japan)* **5**, 738

Tsuzuku, T. and K. Saito (1970) in *Proceedings of the Conference on the Material Science of Carbon and Graphite, Tokyo*

Ubbelohde, A. R. (1959) in *Proceedings of the Third Conference on Carbon*, Pergamon, p. 65

Ubbelohde, A. R. and F. A. Lewis (1960) *Graphite and its Crystal Compounds*, Oxford U.P., London

Vaccari, J. A. (1969) *Materials Engineering* **2**, 36

Wallace, P. R. (1947) *Phys. Rev.* **71**, 622

Warren, B. E. (1941) *Phys. Rev.* **9**, 693

Waters, P. L. (1961) in *Proceedings of the Fifth Conference on Carbon*, Pergamon, Vol. 2, p. 132

Watt, W. (1970) *Proc. Roy. Soc. London A* **319**, 5

Watt, W. (1972) *Carbon* **10**, 121

Watt, W. and W. Johnson (1970) in *Proceedings of the Third Conference on Industrial Carbon and Graphite*, Society of Chemical Industry, London, p. 417

Watt, W., L. N. Phillips and W. Johnson (1966) *The Engineer* **221**, 815

Williams, W. S., D. A. Steffens and R. Bacon (1970) *J. Appl. Phys.* **41**, 4893

Winslow, F. H. (1956) in *Proceedings of the First and Second Conferences on Carbon, Buffalo* (Waverly, Baltimore), p. 93

Winslow, F. H. (1958) in *Proceedings of the Conference on Industrial Carbon and Graphite*, Society of Chemical Industry, London, p. 190

Wolff, W. F. (1958) *J. Phys. Chem.* **62**, 829

Yamada, S. (1967) *Tanso, Japan* **51**, 17

Yamada, S. (1968) *A Review of Glasslike Carbons*, Battelle Memorial Institute, Ohio

Yamada, S. and H. Sato (1962) *Nature* **193**, 261

Yamada, S., H. Sato and T. Ishu (1964) *Carbon* **2**, 253

Yamaguchi, T. (1964) *Carbon* **1**, 47 and 535; *Carbon* **2**, 95

Yokokawa, C., K. Hasokawa and Y. Takagami (1966) *Carbon* **4**, 459

Zinke, A. (1951) *J. Appl. Chem.* **1**, 257

Author Index

Subject Index